Character Tables and Compatibility Relations of The Eighty Layer Groups and Seventeen Plane Groups

Character Tables and Compatibility Relations of The Eighty Layer Groups and Seventeen Plane Groups

Daniel B. Litvin and
Thomas R. Wike
The Pennsylvania State University
The Berks Campus
Reading, Pennsylvania

Plenum Press • New York and London

Library of Congress Cataloging-in-Publication Data

Litvin, Daniel B.
   Character tables and compatibility relations of the eighty layer
groups and seventeen plane groups / Daniel B. Litvin and Thomas R.
Wike.
      p.   cm.
   Includes bibliographical references and index.
   ISBN-13:978-1-4612-7744-6   e-ISBN-13:978-1-4613-0495-1
   DOI: 10.1007/978-1-4613-0495-1
   1. Solid state physics--Tables.   2. Solid state chemistry--Tables.
3. Space groups--Tables.   I. Wike, Thomas R.   II. Title.
QC176.L55   1991
530.4'1'0212--dc20   91-32601
                                                           CIP

ISBN-13:978-1-4612-7744-6

The research in this volume was supported by
the National Science Foundation under Grant No. DMR-8406196
at The Pennsylvania State University

In memory of my grandfather, Baruch Litvin (*zichrono lavracha*),
may I be a "pipefitter." *Toshlaba.*

—DBL

In memory of my father, Elmer E. Wike,
and to my mother, Miriam.

—TRW

CONTENTS

TABLES

CHARACTER TABLES AND COMPATIBILITY RELATIONS
OF THE EIGHTY LAYER GROUPS
AND SEVENTEEN PLANE GROUPS

1. Introduction

The seventeen two-dimensional space groups and the two hundred and thirty three-dimensional space groups are well known. For use in the determination and classification of planar and spatial atom arrangements, extensive tables of the properties of these two-dimensional and three-dimensional space groups are given in a Volume A of the International Tables for Crystallography.[1] For use in labeling electronic bands and lattice vibration spectra, extensive tables of the irreducible representations of the two-dimensional[2-3] and three-dimensional[4-9] space groups have been published.

Of physical interest are additional "layer" atom arrangements whose symmetry is neither one of the seventeen two-dimensional nor one of the two hundred and thirty three-dimensional space groups. These atom arrangements are characterized by two-dimensional translational periodicity and a finite extension in the third dimension, i.e., a finite thickness. The two-dimensional translational periodicity precludes three-dimensional space group symmetry. The finite thickness allows for possible three-dimensional reflections, glide planes, and two-fold screw axes not included in two-dimensional space group symmetry. The symmetry group of such a layer atom arrangement is one of eighty so-called layer groups.

The existence of these eighty layer groups was recognized by several authors (Speiser,[10] Hermann,[11,12] Alexander and Herrmann,[13,14] and Weber[15]) in the late twenties. Other names for these eighty groups are "diperiodic groups"[16] and "two-dimensional (subperiodic) groups in three-dimensional space."[17]

TABLE 1

SCHOENFLIES SYMBOL	INTERNATIONAL	
	SHORT	FULL
C_1	1	1
$C_i(S_2)$	$\bar{1}$	$\bar{1}$
C_2	2	2
$C_s(C_{1h})$	m	m
C_{2h}	2/m	$\frac{2}{m}$
$D_2(V)$	222	222
C_{2v}	mm2	mm2
$D_{2h}(V_h)$	mmm	$\frac{2}{m}\frac{2}{m}\frac{2}{m}$
C_4	4	4
S_4	$\bar{4}$	$\bar{4}$
C_{4h}	4/m	$\frac{4}{m}$
D_4	422	422
C_{4v}	4mm	4mm
$D_{2d}(V_2)$	$\bar{4}$2m	$\bar{4}$2m
D_{4h}	4/mmm	$\frac{4}{m}\frac{2}{m}\frac{2}{m}$
C_3	3	3
$C_{3i}(S_6)$	$\bar{3}$	$\bar{3}$
D_3	32	32
C_{3v}	3m	3m
D_{3d}	$\bar{3}$m	$\bar{3}\frac{2}{m}$
C_6	6	6
C_{3h}	$\bar{6}$	$\bar{6}$
C_{6h}	6/m	$\frac{6}{m}$
D_6	622	622
C_{6v}	6mm	6mm
D_{3h}	$\bar{6}$m2	$\bar{6}$m2
D_{6h}	6/mmm	$\frac{6}{m}\frac{2}{m}\frac{2}{m}$

Crystallographic and physical interest in these layer groups has included the investigation of the structure at the boundary between two parts of a twinned crystal[18-19] and of domain pairs,[20] and the determination of the crystallographic structure of thin films by low-energy electron diffraction.[21,22] In the study of "layered" crystals, the dominance of the layer symmetry with weak interlayer interactions is used to resolve vibrational Davydov splitting.[23-25]

In this work we shall consider the group-theoretical structure of the layer groups and their irreducible representations. In the following sections we shall tabulate the properties of the point groups and translation groups associated with the eighty layer groups. The nomenclature, symbol, and elements of the eighty layer groups are given. The reciprocal lattices and related wave vectors k used in labeling irreducible representations are then tabulated. Finally, we give the character tables and compatibility relations of the irreducible representations of the eighty layer groups.

2. Point Groups

The twenty-seven point groups associated with layer groups are listed in Table 1 in Schoenflies, short international, and full international notation. These are all of the thirty-two point groups associated with the three-dimensional space groups excluding the point groups T, T_h, O, T_d, and O_h. All twenty-seven of the layer point groups are subgroups of the point group D_{4h} or D_{6h}. Consequently, to identify and label the elements of the layer point groups we shall identify and label the elements of the point groups D_{4h} and D_{6h}.

In Figure 1 we give the orthogonal coordinate system, with z-axis perpendicular to the plane of the figure, in which the elements of the point group D_{4h} are defined In Table 2 we identify and label the elements of the point group D_{4h}, providing the

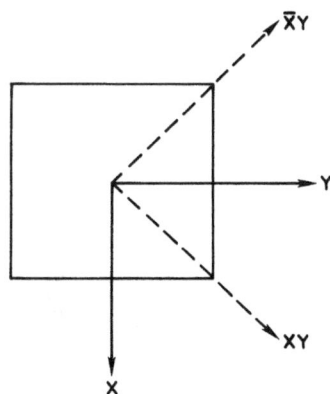

Figure 1. The coordinate system for the point group D_{4h}.

following information:

Column 1:   The proper rotations R of the point group D_{4h} in Schoenflies notation.

Column 2:   The action of the proper rotation R on the point (x,y,z).

Column 3:   The direction cosines of the rotation axis of R with respect to the orthogonal coordinate system of Figure 1, and the angle of rotation.

Column 4:   The Schoenflies notation used by Bradley and Cracknell.[8]

Column 5:   The numerical labeling used by Miller and Love[6] and Cracknell, Davies, Miller, and Love.[9]

Column 6:   The numerical label of the double-valued point group element $\overline{E}R$. The numerical label of $\overline{E}R$ is the numerical label of R given in column 5 plus forty-eight.

Column 7:   The improper rotations IR, i.e., the proper rotations R combined with inversion I, in Schoenflies notation.

Column 8:   The action of the improper rotation IR on the point (x,y,z).

Column 9:   The Schoenflies notation used by Bradley and Cracknell.[8]

Column 10: The Miller and Love[6] and Cracknell, Davies, Miller, and Love[9] numerical label. The numerical label of the improper rotation IR is the numerical label of the proper rotation R given in column 5 plus twenty-four.

Column 11: The numerical label of the double-valued point group element $\overline{E}IR$. The numerical label of $\overline{E}IR$ is the numerical label of IR given in column 10 plus forty-eight.

In this work we shall denote the elements of the point group D_{4h} by the Schoenflies notation of columns 1 and 7 and the numerical labels of columns 5, 6, 10, and 11.

TABLE 2

1	2	3	4	5	6	7	8	9	10	11
C_1	x,y,z		E	1	49	I	-x,-y,-z	I	25	73
C_{2x}	x,-y,-z	(1,0,0);180	C_{2x}	2	50	m_x	-x,y,z	σ_x	26	74
C_{2y}	-x,y,-z	(0,1,0);180	C_{2y}	3	51	m_y	x,-y,z	σ_y	27	75
C_{2z}	-x,-y,z	(0,0,1);180	C_{2z}	4	52	m_z	x,y,-z	σ_z	28	76
$C_{2\overline{xy}}$	-y,-x,-z	(-α,α,0);180	C_{2b}	13	61	$m_{\overline{xy}}$	y,x,z	σ_d	37	85
C_{4z}	-y,x,z	(0,0,1);90	C_{4z}^+	14	62	S_{4z}^{-1}	y,-x,-z	S_{4z}^-	38	86
C_{4z}^{-1}	y,-x,z	(0,0,1);270	C_{4z}^-	15	63	S_{4z}	-y,x,-z	S_{4z}^+	39	87
C_{2xy}	y,x,-z	(α,α,0);180	C_{2a}	16	64	m_{xy}	-y,-x,z	σ_{da}	40	88

$\alpha = 1/\sqrt{2}$ $\cos^{-1}\alpha = 45°$

4

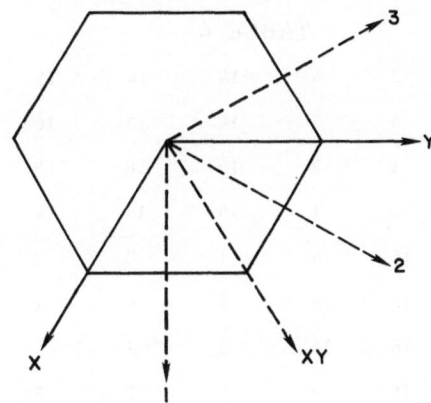

Figure 2. The coordinate system for the point group D_{6h}.

In Figure 2 we give the hexagonal coordinate system in which the elements of the point group D_{6h} are defined. The angle between the x- and y-axes is 120° and the z-axis is perpendicular to the plane of the figure. In Table 3 we identify and label the elements of the point group D_{6h}, each column giving the same information as the corresponding column in Table 2, with the following changes:

Column 3:   The direction cosines are given with respect to the orthogonal coordinate system (x,2,z) of Figure 2.

Column 10: The numerical label of the improper rotation IR is the numerical label of the proper rotation R given in column 5 plus twelve.

TABLE 3

1	2	3	4	5	6	7	8	9	10	11
C_1	x,y,z		E	1	49	I	x,-y,-z	I	13	61
C_6	x-y,x,z	(0,0,1);60	C_6^+	2	50	S_3^{-1}	-x+y,-x,-z	S_3^-	14	62
C_3	-y,x-y,z	(0,0,1);120	C_3^+	3	51	S_6^{-1}	y,-x+y,-z	S_6^-	15	63
C_2	-x,-y,z	(0,0,1);180	C_2	4	52	m_z	x,y,-z	σ_h	16	64
C_3^{-1}	-x+y,-x,z	(0,0,1);240	C_3^-	5	53	S_6	x-y,x,-z	S_6^+	17	65
C_6^{-1}	y,-x+y,z	(0,0,1);300	C_6^-	6	54	S_3	-y,x-y,-z	S_3^+	18	66
C_{2x}	x-y,-y,-z	(1,0,0);180	C_{21}''	7	55	m_x	-x+y,y,z	σ_{v1}	19	67
C_{21}	x,x-y,-z	$(\gamma,\delta,0)$;180	C_{22}'	8	56	m_1	-x,-x+y,z	σ_{d2}	20	68
C_{2xy}	y,x,-z	$(\delta,\gamma,0)$;180	C_{23}''	9	57	m_{xy}	-y,-x,z	σ_{v3}	21	69
C_{22}	-x+y,y,-z	(0,1,0);180	C_{21}'	10	58	m_2	x-y,-y,z	σ_{d1}	22	70
C_{2y}	-x,-x+y,-z	$(-\delta,\gamma,0)$;180	C_{22}''	11	59	m_y	x,x-y,z	σ_{v2}	23	71
C_{23}	-y,-x,-z	$(-\gamma,\delta,0)$;180	C_{23}'	12	60	m_3	y,x,z	σ_{d3}	24	72

$$\gamma = \frac{\sqrt{3}}{2} \quad \cos^{-1}\gamma = 30° \qquad\qquad \delta = \tfrac{1}{2} \quad \cos^{-1}\delta = 60°$$

5

TABLE 4

1	2	3	4	13	14	15	16
2	1	4	3	14	13	16	15
3	4	1	2	15	16	13	14
4	3	2	1	16	15	14	13
13	15	14	16	1	3	2	4
14	16	13	15	2	4	1	3
15	13	16	14	3	1	4	2
16	14	15	13	4	2	3	1

In this work we shall denote the elements of the point group D_{6h} in the same manner as the elements of D_{4h}, by the Schoenflies notation of columns 1 and 7 and the numerical labels of column 5, 6, 10, and 11.

Information to construct the multiplication tables of both the "single-" and "double-valued" point groups D_{4h} and D_{6h} is given in Tables 4-11. For "single-valued" point groups, the multiplication table of the proper rotations of each of the point groups D_{4h} and D_{6h} is given explicitly in Tables 4 and 5, respectively. The entry at the intersection of the ith row and jth column is the product $R_i R_j$ of the element R_i in the first column of the ith row and the element R_j in the first row of the jth column. For example, from Table 4, 3 x 15 = 13. Tables 4 and 5 provide,

TABLE 5

1	2	3	4	5	6	7	8	9	10	11	12
2	3	4	5	6	1	8	9	10	11	12	7
3	4	5	6	1	2	9	10	11	12	7	8
4	5	6	1	2	3	10	11	12	7	8	9
5	6	1	2	3	4	11	12	7	8	9	10
6	1	2	3	4	5	12	7	8	9	10	11
7	12	11	10	9	8	1	6	5	4	3	2
8	7	12	11	10	9	2	1	6	5	4	3
9	8	7	12	11	10	3	2	1	6	5	4
10	9	8	7	12	11	4	3	2	1	6	5
11	10	9	8	7	12	5	4	3	2	1	6
12	11	10	9	8	7	6	5	4	3	2	1

TABLE 6   TABLE 7

TABLE 6

	R	IR
R	N	N+24
IR	N+24	N

TABLE 7

	R	IR
R	N	N+12
IR	N+12	N

respectively, only one quarter of the multiplication tables of the point groups D_{4h} and D_{6h}. The complete multiplication tables are given schematically in Tables 6 and 7. For example, for the point group D_{4h}, if the numerical label of the product of $R_i R_j$ is N, then the numerical label of $R_i(IR_j)$ is N + 24. Consequently the entry in the upper right block of Table 6 is N + 24.

For the "double-valued" multiplication tables of the point groups D_{4h} and D_{6h}, the multiplication tables of the proper rotations are given, respectively, in Tables 8 and 9. The complete multiplication tables are given schematically in Tables 10 and 11. The symbol N ± 48 takes the value N + 48 if N < 48 and N - 48 if > 48.

The twenty-seven layer point groups are subgroups of the point groups D_{4h} and D_{6h}. The elements of the point groups D_{4h} and D_{6h} and each of their subgroups are given explicitly in Tables 12 and 13, respectively. If two or more subgroups belong to the same class of point groups, notations have been added in superscript to the point group symbol to uniquely label each subgroup. For example, in Table 12 one finds listed the two subgroups $D_2^{(x,y,z)}$ and $D_2^{(xy,\bar{x}y,z)}$ of D_{4h}. The subgroup

TABLE 8

1	2	3	4	13	14	15	16
2	49	4	51	14	61	64	15
3	52	49	2	15	16	61	62
4	3	50	49	64	15	62	13
13	63	62	16	49	3	2	52
14	16	13	15	50	4	49	3
15	13	64	62	51	49	52	50
16	62	15	61	4	2	51	49

TABLE 9

1	2	3	4	5	6	7	8	9	10	11	12
2	3	4	5	6	49	8	9	10	11	12	55
3	4	5	6	49	50	9	10	11	12	55	56
4	5	6	49	50	51	10	11	12	55	56	57
5	6	49	50	51	52	11	12	55	56	57	58
6	49	50	51	52	53	12	55	56	57	58	59
7	60	59	58	57	56	49	6	5	4	3	2
8	7	60	59	58	57	50	49	6	5	4	3
9	8	7	60	59	58	51	50	49	6	5	4
10	9	8	7	60	59	52	51	50	49	6	5
11	10	9	8	7	60	53	52	51	50	49	6
12	11	10	9	8	7	54	53	52	51	50	49

relations among the point groups D_{4h} and D_{6h} and the classes of their subgroups are given diagrammatically in Figures 3 and 4. The subgroup relations are given explicitly in Tables 14 and 15. An "X" at the intersection of the ith row and jth column denotes that the point group labeling the ith row is a subgroup of the point group labeling the jth column.

3. Translation Groups

The five classes of translation groups associated with the eighty layer groups are the five classes of two-dimensional translation groups. The generating translations of representatives of these five translation groups are shown in Figure 5. In Table 16 we list the nomenclature of these five translation groups and the con-

TABLE 10

	R	IR	$\bar{E}R$	$\bar{E}IR$
R	N	N+24	N±48	N+24±48
IR	N+24	N	N+24±48	N±48
$\bar{E}R$	N±48	N+24±48	N	N+24
$\bar{E}IR$	N+24±48	N±48	N+24	N

TABLE 11

	R	IR	$\bar{E}R$	$\bar{E}IR$
R	N	N+12	N±48	N+12±48
IR	N+12	N	N+12±48	N±48
$\bar{E}R$	N±48	N+12±48	N	N+12
$\bar{E}IR$	N+12±48	N±48	N+12	N

TABLE 12

	C_1	C_{2x}	C_{2y}	C_{2z}	$C_{2\bar{x}y}$	C_{2xy}	C_{4z}	C_{4z}^{-1}
D_{4h}	I	m_x	m_y	m_z	$m_{\bar{x}y}$	m_{xy}	S_{4z}	S_{4z}^{-1}
C_{4h}	C_1	C_{4z}	C_{2z}	C_{4z}^{-1}	I	S_{4z}	m_z	S_{4z}^{-}
D_4	C_1	C_{4z}	C_{2z}	C_{4z}^{-1}	C_{2x}	C_{2y}	C_{2xy}	$C_{2\bar{x}y}$
$D_{2d}^{(x,y,z)}$	C_1	S_{4z}	C_{2z}	S_{4z}^{-1}	C_{2x}	C_{2y}	m_{xy}	$m_{\bar{x}y}$
$D_{2d}^{(xy,\bar{x}y,z)}$	C_1	S_{4z}	C_{2z}	S_{4z}^{-1}	m_x	m_y	C_{2xy}	$C_{2\bar{x}y}$
C_{4v}	C_1	C_{4z}	C_{2z}	C_{4z}^{-1}	m_x	m_y	$m_{\bar{x}y}$	m_{xy}
$D_{2h}^{(x,y,z)}$	C_1	C_{2x}	C_{2y}	C_{2z}	I	m_x	m_y	m_z
$D_{2h}^{(\bar{x}y,xy,z)}$	C_1	C_{2xy}	$C_{2\bar{x}y}$	C_{2z}	1	m_{xy}	$m_{\bar{x}y}$	m_z
S_4	C_1	S_{4z}	C_{2z}	S_{4z}^{-1}				
C_4	C_1	C_{4z}	C_{2z}	C_{4z}^{-1}				
$D_2^{(x,y,z)}$	C_1	C_{2x}	C_{2y}	C_{2z}				
$D_2^{(xy,\bar{x}y,z)}$	C_1	C_{2xy}	$C_{2\bar{x}y}$	C_{2z}				
$C_{2h}^{(z)}$	C_1	C_{2z}	I	m_z				
$C_{2h}^{(x)}$	C_1	C_{2x}	I	m_x				
$C_{2h}^{(y)}$	C_1	C_{2y}	I	m_y				
$C_{2h}^{(xy)}$	C_1	C_{2xy}	I	m_{xy}				
$C_{2h}^{(\bar{x}y)}$	C_1	$C_{2\bar{x}y}$	I	$m_{\bar{x}y}$				
$C_{2v}^{(z,x,y)}$	C_1	C_{2z}	m_x	m_y				
$C_{2v}^{(z,xy,\bar{x}y)}$	C_1	C_{2z}	m_{xy}	$m_{\bar{x}y}$				
$C_{2v}^{(x)}$	C_1	C_{2x}	m_y	m_z				
$C_{2v}^{(y)}$	C_1	C_{2y}	m_x	m_z				
$C_{2v}^{(xy)}$	C_1	C_{2xy}	$m_{\bar{x}y}$	m_z				
$C_{2v}^{(\bar{x}y)}$	C_1	$C_{2\bar{x}y}$	m_{xy}	m_z				

(continued)

TABLE 12
(continued)

C_1	C_1	I
$C_2^{(z)}$	C_1	C_{2z}
$C_2^{(x)}$	C_1	C_{2x}
$C_2^{(y)}$	C_1	C_{2y}
$C_2^{(xy)}$	C_1	C_{2xy}
$C_2^{(\bar{x}y)}$	C_1	$C_{2\bar{x}y}$
$C_s^{(x)}$	C_1	m_x
$C_s^{(y)}$	C_1	m_y
$C_s^{(z)}$	C_1	m_z
$C_s^{(xy)}$	C_1	m_{xy}
$C_s^{(\bar{x}y)}$	C_1	$m_{\bar{x}y}$
C_1	C_1	

TABLE 13

D_{6h}	C_1	C_6	C_3	C_2	C_3^{-1}	C_6^{-1}
	C_{2x}	C_{2y}	C_{2xy}	C_{21}	C_{22}	C_{23}
	I	S_3^{-1}	S_6^{-1}	m_z	S_6	S_3
	m_x	m_y	m_{xy}	m_1	m_2	m_3
D_6	C_1	C_6	C_3	C_2	C_3^{-1}	C_6^{-1}
	C_{2x}	C_{2y}	C_{2xy}	C_{21}	C_{22}	C_{23}
C_{6h}	C_1	C_6	C_3	C_2	C_3^{-1}	C_6^{-1}
	I	C_3^{-1}	S_6^{-1}	m_z	S_6	S_3
C_{6v}	C_1	C_6	C_3	C_2	C_3^{-1}	C_6^{-1}
	m_x	m_y	m_{xy}	m_1	m_2	m_3

(continued)

TABLE 13
(continued)

$D_{3d}^{(x,y,xy)}$	c_1	c_3	c_3^{-1}	m_x	m_y	m_{xy}
	I	s_6^{-1}	s_6	c_{2x}	c_{2y}	c_{2xy}
$D_{3d}^{(1,2,3)}$	c_1	c_3	c_3^{-1}	m_1	m_2	m_3
	I	s_6^{-1}	s_6	c_{21}	c_{22}	c_{23}
$D_{3h}^{(x,y,xy)}$	c_1	c_3	c_3^{-1}	c_{2x}	c_{2y}	c_{2xy}
	m_z	s_3	s_3^{-1}	m_1	m_2	m_3
$D_{3h}^{(1,2,3)}$	c_1	c_3	c_3^{-1}	c_{21}	c_{22}	c_{23}
	m_z	s_3	s_3^{-1}	m_x	m_y	m_{xy}
$D_{2h}^{(x,2,z)}$	c_1	c_{2x}	c_{22}	c_{2z}		
	I	m_x	m_2	m_z		
$D_{2h}^{(y,1,z)}$	c_1	c_{21}	c_{2y}	c_{2z}		
	I	m_1	m_y	m_z		
$D_{2h}^{(xy,3,z)}$	c_1	c_{2xy}	c_{23}	c_{2z}		
	I	m_{xy}	m_3	m_z		
c_6	c_1	c_6	c_3	c_2	c_3^{-1}	c_6^{-1}
c_{3i}	c_1	c_3	c_3^{-1}	I	s_6^{-1}	s_6
$D_3^{(x,y,xy)}$	c_1	c_3	c_3^{-1}	c_{2x}	c_{2y}	c_{2xy}
$D_3^{(1,2,3)}$	c_1	c_3	c_3^{-1}	c_{21}	c_{22}	c_{23}
$c_{3v}^{(x,y,xy)}$	c_1	c_3	c_3^{-1}	m_x	m_y	m_{xy}
$c_{3v}^{(1,2,3)}$	c_1	c_3	c_3^{-1}	m_1	m_2	m_3
c_{3h}	c_1	c_3	c_3^{-1}	m_z	s_3	s_3^{-1}

(continued)

11

TABLE 13
(continued)

$D_2^{(x,2,z)}$	C_1	C_{2x}	C_{22}	C_{2z}
$D_2^{(y,1,z)}$	C_1	C_{21}	C_{2y}	C_{2z}
$D_2^{(xy,3,z)}$	C_1	C_{2xy}	C_{23}	C_{2z}
$C_{2h}^{(z)}$	C_1	C_{2z}	I	m_z
$C_{2h}^{(x)}$	C_1	C_{2x}	I	m_x
$C_{2h}^{(y)}$	C_1	C_{2y}	I	m_y
$C_{2h}^{(xy)}$	C_1	C_{2xy}	I	m_{xy}
$C_{2h}^{(1)}$	C_1	C_{21}	I	m_1
$C_{2h}^{(2)}$	C_1	C_{22}	I	m_2
$C_{2h}^{(3)}$	C_1	C_{23}	I	m_3
$C_{2v}^{(z,x,2)}$	C_1	C_{2z}	m_x	m_2
$C_{2v}^{(z,y,1)}$	C_1	C_{2z}	m_y	m_1
$C_{2v}^{(z,xy,3)}$	C_1	C_{2z}	m_{xy}	m_3
$C_{2v}^{(x)}$	C_1	C_{2x}	m_2	m_z
$C_{2v}^{(y)}$	C_1	C_{2y}	m_1	m_z
$C_{2v}^{(xy)}$	C_1	C_{2xy}	m_3	m_z
$C_{2v}^{(1)}$	C_1	C_{21}	m_y	m_z
$C_{2v}^{(2)}$	C_1	C_{22}	m_x	m_z

(continued)

TABLE 13
(continued)

$c_{2v}^{(3)}$	c_1	c_{23}	m_{xy}	m_z
c_3	c_1	c_3	c_3^{-1}	
c_i	c_1	I		
$c_2^{(z)}$	c_1	c_{2z}		
$c_2^{(x)}$	c_1	c_{2x}		
$c_2^{(y)}$	c_1	c_{2y}		
$c_2^{(xy)}$	c_1	c_{2xy}		
$c_2^{(1)}$	c_1	c_{21}		
$c_2^{(2)}$	c_1	c_{22}		
$c_2^{(3)}$	c_1	c_{23}		
$c_s^{(z)}$	c_1	m_z		
$c_s^{(x)}$	c_1	m_x		
$c_s^{(y)}$	c_1	m_y		
$c_s^{(xy)}$	c_1	m_{xy}		
$c_s^{(1)}$	c_1	m_1		
$c_s^{(2)}$	c_1	m_2		
$c_s^{(3)}$	c_1	m_3		
c_1	c_1			

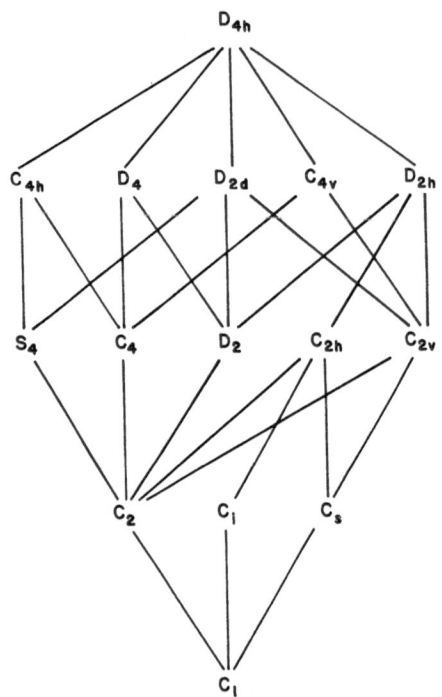

Figure 3.   Subgroups of D_{4h}.

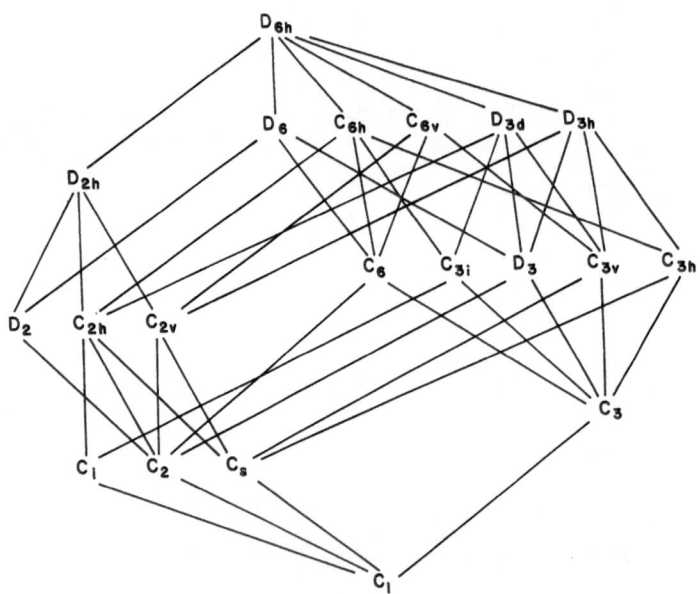

Figure 4.   Subgroups of D_{6h}.

TABLE 14

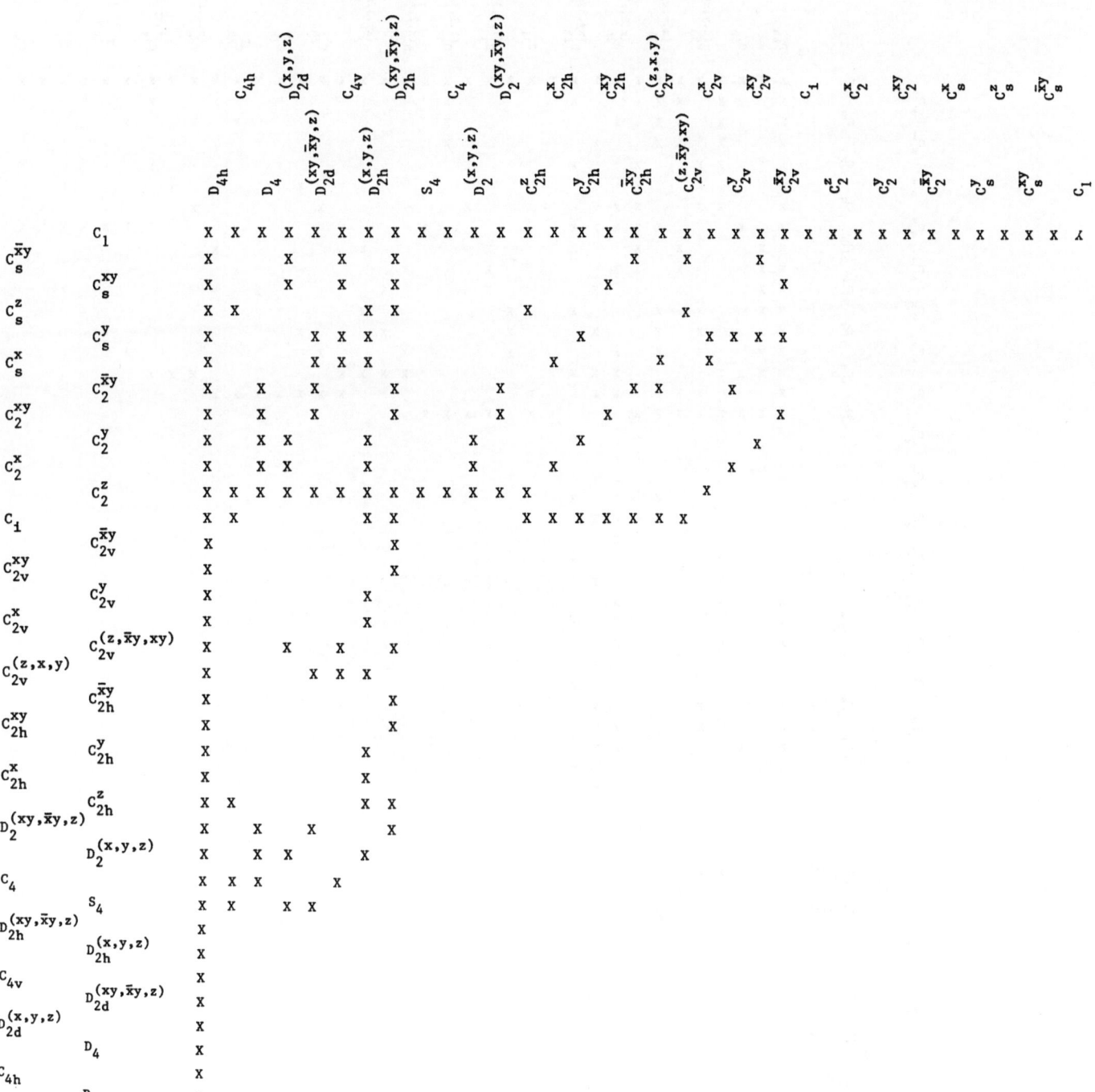

TABLE 15

		D_6	D_{6h}	C_{6v}	C_{6h}	$D_{3d}^{(1,2,3)}$	$D_{3d}^{(x,y,xy)}$	$D_{3h}^{(1,2,3)}$	$D_{3h}^{(x,y,xy)}$	$D_{2h}^{(y,1,z)}$	$D_{2h}^{(x,2,z)}$	$D_{2h}^{(xy,3,z)}$	C_6	C_{3i}	$D_3^{(x,y,xy)}$	$D_3^{(1,2,3)}$	$C_{3v}^{(x,y,xy)}$	$C_{3v}^{(1,2,3)}$	C_{3h}	$D_2^{(y,1,z)}$	$D_2^{(x,2,z)}$	$D_2^{(xy,3,z)}$	C_{2h}^z	C_{2h}^x	C_{2h}^y	C_{2h}^{xy}	C_{2h}^1	C_{2h}^2	C_{2h}^3	$C_{2v}^{(z,y,1)}$	$C_{2v}^{(z,x,2)}$	C_{2v}^x	$C_{2v}^{(z,xy,3)}$	C_{2v}^{xy}	C_{2v}^y	C_{2v}^1	C_{2v}^2	C_{2v}^3
	C_1	X	X	X	X	X	X	X	X	X	X	X	X	X	X	X	X	X	X	X	X	X	X	X	X	X	X	X	X	X	X	X	X	X	X	X	X	X
C_s^3	C_s^2	X		X		X	X			X									X										X			X				X		
C_s^1		X		X		X	X		X										X									X				X			X			
	C_s^{xy}	X		X	X		X		X		X			X							X			X							X			X				X
C_s^y	C_s^x	X		X	X		X		X	X			X								X				X								X			X		X
C_s^z		X	X			X	X	X	X	X	X								X					X				X		X	X	X	X	X	X	X	X	X
C_2^2	C_2^3	X	X			X		X	X				X				X				X			X					X				X				X	
C_2^1		X	X			X		X	X				X				X				X				X				X					X			X	
C_2^{xy}	C_2^y	X	X			X	X			X			X			X				X			X			X					X		X			X		
C_2^x		X	X			X	X		X	X			X		X						X			X			X			X				X				
	C_2^z	X	X	X	X			X	X	X	X	X								X	X	X	X							X	X	X						
C_1		X		X	X			X	X	X			X	X									X	X	X	X	X	X	X									
	C_3	X	X	X	X	X	X	X	X				X	X	X	X	X	X	X																			
C_{2v}^3	C_{2v}^2	X				X			X																													
C_{2v}^1		X				X	X																															
C_{2v}^y	C_{2v}^{xy}	X					X		X																													
	C_{2v}^x	X					X	X																														
$C_{2v}^{(z,xy,3)}$	$C_{2v}^{(z,y,1)}$	X		X					X																													
$C_{2v}^{(z,x,2)}$		X		X				X																														
C_{2h}^2	C_{2h}^3	X			X			X												X																		
C_{2h}^{xy}	C_{2h}^1	X			X			X		X										X																		
	C_{2h}^{xy}	X				X			X																													
C_{2h}^x	C_{2h}^y	X				X			X																													
	C_{2h}^x	X				X			X																													
$D_2^{(xy,3,z)}$	C_{2h}^z	X	X				X	X	X																													
$D_2^{(x,2,z)}$	$D_2^{(y,1,z)}$	X	X					X																														
	$D_2^{(x,2,z)}$	X	X					X																														
$C_{3v}^{(1,2,3)}$	C_{3h}	X		X			X	X																														
$D_3^{(1,2,3)}$	$C_{3v}^{(x,y,xy)}$	X		X	X		X		X																													
	$D_3^{(x,y,xy)}$	X	X			X	X																															
C_{3i}		X		X	X	X	X																															
	C_6	X	X	X																																		
$D_{2h}^{(xy,3,z)}$	$D_{2h}^{(y,1,z)}$	X																																				
$D_{2h}^{(x,2,z)}$		X																																				
	$D_{3h}^{(1,2,3)}$	X																																				
$D_{3h}^{(x,y,xy)}$		X																																				
	$D_{3d}^{(1,2,3)}$	X																																				
$D_{3d}^{(x,y,xy)}$		X																																				
	C_{6v}	X																																				
C_{6h}		X																																				
	D_6	X																																				
D_{6h}		X																																				

TABLE 15 (continued)

C_3 C_2^z C_2^y C_2^{-1} C_2^3 C_s^x C_s^{xy} C_s^2 C_1

C_1 C_2^x C_2^{xy} C_2^2 C_s^z C_s^y C_s^{-1} C_s^3

x x x x x x x x x x x x x x x x x x

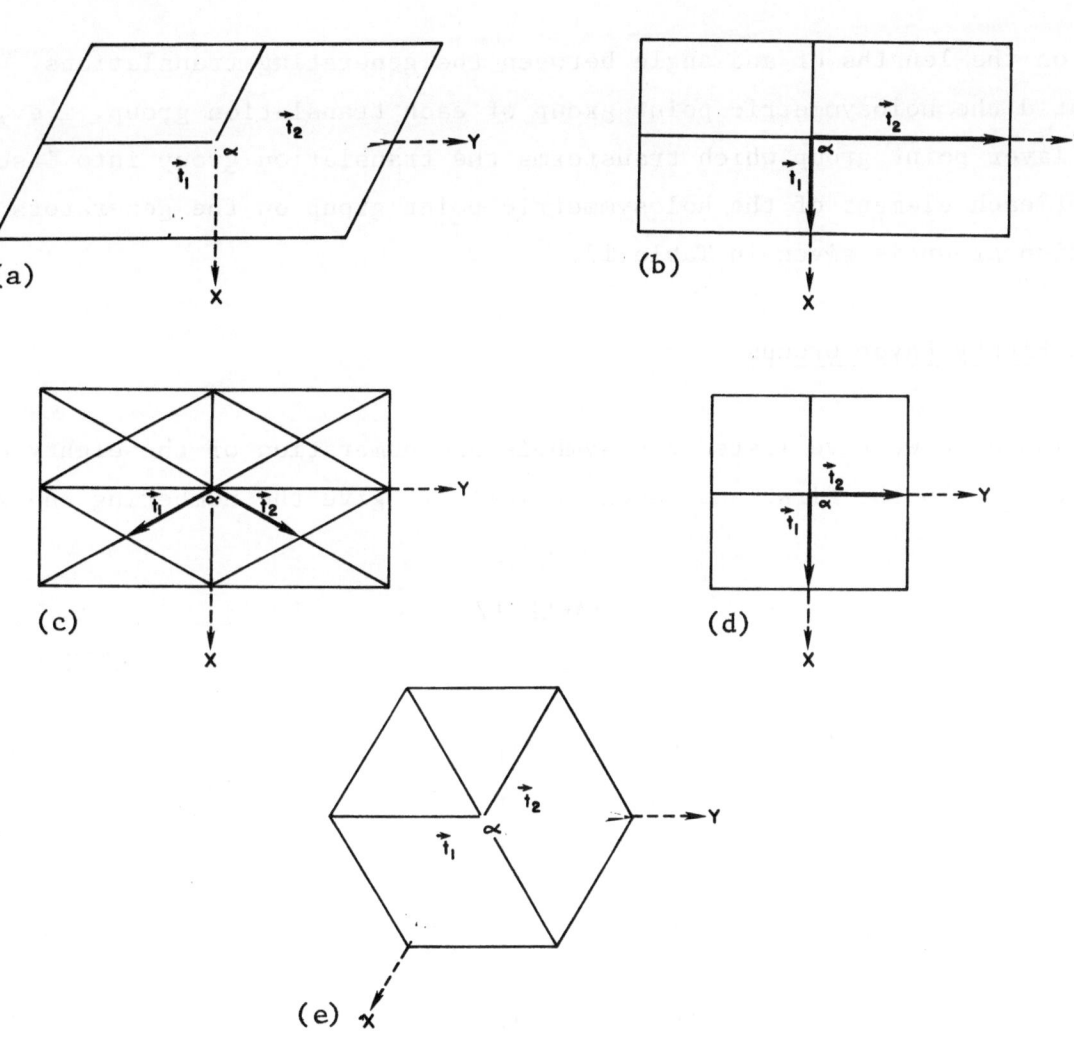

Figure 5. Generators of the two-dimensional translation groups:
(a) Oblique p, (b) Rectangular p, (c) Rectangular c,
(d) Square p, and (e) Hexagonal p.

TABLE 16

1) Oblique p	$\lvert t_1 \rvert \neq \lvert t_2 \rvert$	$\alpha \neq 90°$	C_2
2) Rectangular p	$\lvert t_1 \rvert \neq \lvert t_2 \rvert$	$\alpha = 90°$	D_{2h}
3) Rectangular c	$\lvert t_1 \rvert = \lvert t_2 \rvert$	$\alpha \neq 60°,90°,120°$	D_{2h}
4) Square p	$\lvert t_1 \rvert = \lvert t_2 \rvert$	$\alpha = 90°$	D_{4h}
5) Hexagonal p	$\lvert t_1 \rvert = \lvert t_2 \rvert$	$\alpha = 120°$	D_{6h}

ditions on the lengths of and angle between the generating translations. We have also listed the holosymmetric point group of each translation group, i.e., the largest layer point group which transforms the translation group into itself. The action of each element of the holosymmetric point group on the generators of each translation group is given in Table 17.

4. The Eighty Layer Groups

In Table 18 we have listed the symbols and numeration of the eighty classes of layer groups. In the first and second columns we give the numbering and symbol used

TABLE 17

OBLIQUE p					RECTANGULAR p					RECTANGULAR c			
C_1	1	t_1	t_2		C_1	1	t_1	t_2		C_1	1	t_1	t_2
C_{2z}	4	$-t_1$	$-t_2$		C_{2x}	2	t_1	$-t_2$		C_{2x}	2	t_2	t_1
I	25	$-t_1$	$-t_2$		C_{2y}	3	$-t_1$	t_2		C_{2y}	3	$-t_2$	$-t_1$
m_z	28	t_1	t_2		C_{2z}	4	$-t_1$	$-t_2$		C_{2z}	4	$-t_1$	$-t_2$
					I	25	$-t_1$	$-t_2$		I	25	$-t_1$	$-t_2$
					m_x	26	$-t_1$	t_2		m_x	26	$-t_2$	$-t_1$
					m_y	27	t_1	$-t_2$		m_y	27	t_2	t_1
					m_z	28	t_1	t_2		m_z	28	t_1	t_2

(continued)

TABLE 17
(continued)

SQUARE p

C_1	1	t_1	t_2
C_{2x}	2	t_1	$-t_2$
C_{2y}	3	$-t_1$	t_2
C_{2z}	4	$-t_1$	$-t_2$
$C_{2\overline{xy}}$	13	$-t_2$	$-t_1$
C_{4z}	14	t_2	$-t_1$
C_{4z}^{-1}	15	$-t_2$	t_1
C_{2xy}	16	t_2	t_1
I	25	$-t_1$	$-t_2$
m_x	26	$-t_1$	t_2
m_y	27	t_1	$-t_2$
m_z	28	t_1	t_2
$m_{\overline{xy}}$	37	t_2	t_1
S_{4z}^{-1}	38	$-t_2$	t_1
S_{4z}	39	t_2	$-t_1$
m_{xy}	40	$-t_2$	$-t_1$

HEXAGONAL P

C_1	1	t_1	t_2	I	13	$-t_1$	$-t_2$
C_6	2	t_1+t_2	$-t_1$	S_3^{-1}	14	$-t_1-t_2$	t_1
C_3	3	t_2	$-t_1-t_2$	S_6^{-1}	15	$-t_2$	t_1+t_2
C_2	4	$-t_1$	$-t_2$	m_z	16	t_1	t_2
C_3^{-1}	5	$-t_1-t_2$	t_1	S_6	17	t_1+t_2	$-t_1$
C_6^{-1}	6	$-t_2$	t_1+t_2	S_3	18	t_2	$-t_1-t_2$
C_{2x}	7	t_1	$-t_1-t_2$	m_x	19	$-t_1$	t_1+t_2
C_{21}	8	t_1+t_2	$-t_2$	m_1	20	$-t_1-t_2$	t_2
C_{2xy}	9	t_2	t_1	m_{xy}	21	$-t_2$	$-t_1$
C_{22}	10	$-t_1$	t_1+t_2	m_2	22	t_1	$-t_1-t_2$
C_{2y}	11	$-t_1-t_2$	t_2	m_y	23	t_1+t_2	$-t_2$
C_{23}	12	$-t_2$	$-t_1$	m_3	24	t_2	t_1

TABLE 18

	WOOD	WEBER	HERMANN	ALEXANDER AND HERRMANN	NIGGLI	BOHM AND DORNBERGER- SCHIFF		SHUBNIKOV AND KOPTSIK	
OBLIQUE									
1	p1	1	$C_1\bar{p}$	C_1^1	1P1	1	P11(1)	1	$(a/b)\cdot 1$
2	p1̄	2	$S_2\bar{p}$	C_i^1	1P1̄	2	P1̄1̄(1̄)	3	$(a/b)\cdot\bar{1}$
3	p211	8	$C_2\bar{p}$	C_2^1	1P2	9	P11(2)	5	$(a/b){:}2$
4	pm11	3	$C_{1h}\bar{p}\mu$	C_{1h}^1	mP1	4	P11(m)	2	$(a/b)\cdot m$
5	pb11	4	$C_{1h}\bar{p}\alpha$	C_{1h}^2	aP1	6	P11(b)	4	$(a/b)\cdot\breve{b}$
6	p2/m11	12	$C_{2h}\bar{p}\mu$	C_{2h}^1	mP2	13	P11($\frac{2}{m}$)	6	$(a/b)\cdot m{:}2$
7	p2/b11	13	$C_{2h}\bar{p}\alpha$	C_{2h}^2	aP2	17	P11($\frac{2}{b}$)	7	$(a/b)\cdot\breve{b}{:}2$
RECTANGULAR									
8	p112	9	$D_1\bar{p}1$	C_2^2	1P12	8	P12(1)	14	$(a{:}b)\cdot 2$
9	p112$_1$	10	$D_1\bar{p}2$	C_2^3	1P12$_1$	10	P12$_1$(1)	15	$(a{:}b)\cdot 2_1$
10	c112	11	$D_1\bar{c}1$	C_2^4	1C12	11	C12(1)	16	$(\frac{a+b}{2}/a{:}b)\cdot 2$
11	p11m	5	$C_{1v}\bar{p}\mu$	C_{1h}^3	1P1m	3	P1m(1)	8	$(a{:}b){:}m$
12	p11a	6	$C_{1v}\bar{p}\beta$	C_{1h}^4	1P1g	5	P1a(1)	10	$(a{:}b){:}\tilde{a}$

(continued)

TABLE 18
(continued)

	WOOD	WEBER	HERMANN	ALEXANDER AND HERRMANN	NIGGLI	BOHM AND DORNBERGER-SCHIFF	SHUBNIKOV AND KOPTSIK
13	c11m	7	$C_{1v}\bar{c}\mu$	C_{1h}^5	1C1m	7 C1m(1)	12 $(\frac{a+b}{2}/a{:}b){:}m$
14	p112/m	14	$D_{1d}\bar{p}\mu 1$	C_{2h}^3	1P12/m	12 $P1\frac{2}{m}(1)$	17 $(a{:}b)\cdot 2{:}m$
15	p112_1/m	15	$D_{1d}\bar{p}\mu 2$	C_{2h}^5	1P12_1/m	14 $P1\frac{2}{m}1(1)$	18 $(a{:}b)\cdot 2_1{:}m$
16	c112/m	16	$D_{1d}\bar{c}\mu 1$	C_{2h}^7	1C12/m	15 $C1\frac{2}{m}(1)$	19 $(\frac{a+b}{2}/a{:}b)\cdot 2{:}m$
17	p112/a	18	$D_{1d}\bar{p}\beta 2$	C_{2h}^6	1P12/g	16 $P1\frac{2}{a}(1)$	20 $(a{:}b)\cdot 2\cdot\tilde{a}$
18	p112_1/a	17	$D_{1d}\bar{p}\beta 1$	C_{2h}^4	1P12_1/g	18 $P1\frac{2}{a}1(1)$	21 $(a{:}b)\cdot 2_1{:}\tilde{a}$
19	p222	33	$D_2\bar{p}11$	V^1	1P222	33 P22(2)	37 $(a{:}b){:}2{:}2$
20	p222_1	34	$D_2\bar{p}12$	V^3	1P222_1	34 $P2_12(2)$	38 $(a{:}b){:}2{:}2_1$
21	p22_12_1	35	$D_2\bar{p}22$	V^2	1P22_12_1	35 $P2_12_1(2)$	39 $(a{:}b){:}2_1{:}2_1$
22	c222	36	$D_2\bar{c}11$	V^4	1C222	36 C22(2)	40 $(\frac{a+b}{2}/a{:}b){:}2{:}2$
23	p2mm	19	$C_{2v}\bar{p}\mu\mu$	C_{2v}^1	1P2mm	19 Pmm(2)	22 $(a{:}b){:}2\cdot m$
24	pmm2	23	$D_{1h}\bar{p}\mu\mu$	C_{2v}^4	mP12m	20 P2m(m)	9 $(a{:}b)\cdot m\cdot 2$
25	pm2_1a	24	$D_{1h}\bar{p}\mu\beta$	C_{2v}^7	mP12_1g	22 $P2_1a(m)$	11 $(a{:}b)\cdot m\cdot 2_1$
26	pbm2_1	25	$D_{1h}\bar{p}\beta\mu$	C_{2v}^5	aP12_1m	21 $P2_1m(a)$	30 $(a{:}b){:}m\cdot 2_1$
27	pbb2	26	$D_{1h}\bar{p}\beta\beta$	C_{2v}^6	aP12g	23 P2a(a)	31 $(a{:}b)\cdot\tilde{a}\cdot 2$
28	p2ma	20	$C_{2v}\bar{p}\mu\alpha$	C_{2v}^2	1P2mg	24 Pma(2)	24 $(a{:}b){:}2\cdot\tilde{b}$
29	pam2	27	$D_{1h}\bar{p}\alpha\mu$	C_{2v}^{11}	bP12m	25 P2m(b)	32 $(a{:}b)\cdot\tilde{b}\cdot 2$
30	pab2_1	28	$D_{1h}\bar{p}\alpha\beta$	C_{2v}^{14}	bP12_1g	26 $P2_1a(b)$	33 $(a{:}b)\cdot\tilde{b}{:}\tilde{a}$
31	pnb2	29	$D_{1h}\bar{p}\nu\beta$	C_{2v}^{12}	nP12g	27 P2a(n)	34 $(a{:}b)\cdot\widetilde{ab}\cdot 2$
32	pnm2_1	30	$D_{1h}\bar{p}\nu\mu$	C_{2v}^{13}	nP12_1m	28 $P2_1m(n)$	35 $(a{:}b)\cdot\widetilde{ab}\cdot 2_1$
33	p2ba	21	$C_{2v}\bar{p}\beta\alpha$	C_{2v}^{10}	1P2gg	29 Pba(2)	26 $(a{:}b){:}\tilde{a}{:}\tilde{b}$
34	c2mm	22	$C_{2v}\bar{c}\mu\mu$	C_{2v}^3	1C2mm	30 Cmm(2)	28 $(\frac{a+b}{2}/a{:}b){:}m\cdot 2$
35	cmm2	31	$D_{1h}\bar{c}\mu\mu$	C_{2v}^8	mC12m	31 C2m(m)	13 $(\frac{a+b}{2}/a{:}b)\cdot m\cdot 2$
36	cam2	32	$D_{1h}\bar{c}\alpha\mu$	C_{2v}^9	aC12m	32 Cm2(a)	36 $(\frac{a+b}{2}/a{:}b)\cdot\tilde{b}\cdot 2$
37	p2/m2/m2/m	37	$D_{2h}\bar{p}\mu\mu\mu$	V_h^1	mP2mm	37 $P\frac{22}{mm}(\frac{2}{m})$	23 $(a{:}b)\cdot m{:}2\cdot m$
38	p2/a2/m2/a	38	$D_{2h}\bar{p}\alpha\mu\alpha$	V_h^5	aP2mg	38 $P\frac{22}{ma}(\frac{2}{a})$	41 $(a{:}b)\cdot\tilde{a}{:}2\cdot\tilde{a}$
39	p2/n2/b2/a	39	$D_{2h}\bar{p}\nu\beta\alpha$	V_h^6	nP2gg	39 $P\frac{22}{ba}(\frac{2}{n})$	42 $(a{:}b)\cdot\widetilde{ab}{:}2\cdot a$
40	p2/m2_1/m2/a	40	$D_{2h}\bar{p}\mu\mu\alpha$	V_h^3	mP2mg	41 $P\frac{22}{bm}1(\frac{2}{m})$	25 $(a{:}b)\cdot m{:}2\cdot\tilde{b}$
41	p2/a2_1/m2/m	41	$D_2\bar{p}\alpha\mu\mu$	V_h^9	aP2mm	40 $P\frac{2}{m}\frac{2}{m}(\frac{2}{a})$	43 $(a{:}b)\cdot\tilde{a}{:}2\cdot m$
42	p2/n2/m2_1/a	42	$D_{2h}\bar{p}\nu\mu\alpha$	V_h^{11}	nP2mg	42 $P\frac{2}{b}\frac{2}{m}(\frac{2}{a})$	44 $(a{:}b)\cdot\widetilde{ab}{:}2\cdot\tilde{b}$
43	p2/a2/b2_1/a	43	$D_{2h}\bar{p}\alpha\beta\alpha$	V_h^{10}	aP2gg	43 $P\frac{22}{ba}1(\frac{2}{a})$	45 $(a{:}b)\cdot\tilde{a}\cdot 2{:}\tilde{b}$
44	p2/m2_1/b2_1/a	44	$D_{2h}\bar{p}\mu\beta\alpha$	V_h^2	mP2gg	44 $P\frac{2}{b}\frac{2}{a}(\frac{2}{m})$	27 $(a{:}b)\cdot m{:}\tilde{a}{:}\tilde{b}$

(continued)

TABLE 18
(continued)

	WOOD	WEBER	HERMANN	ALEXANDER AND HERRMANN	NIGGLI	BOHM AND DORNBERGER-SCHIFF	SHUBNIKOV AND KOPTSIK
45	p2/a2$_1$/b2$_1$/m	45	$D_{2h}\bar{p}\alpha\beta\mu$	V_h^7	aP2gm	45 $P\frac{2}{m}1\frac{2}{a}1(\frac{2}{b})$	46 (a:b)·$\tilde{b}$:2·$\bar{a}$
46	p2/n2$_1$/m2$_1$/m	46	$D_{2h}\bar{p}\nu\mu\mu$	V_h^8	nP2mm	46 $P\frac{2}{m}1\frac{2}{m}1(\frac{2}{n})$	47 (a:b)·$\widehat{ab}$:2·m
47	c2/m2/m2/m	47	$D_{2h}\bar{c}\mu\mu\mu$	V_h^4	mC2mm	47 $C\frac{2}{mm}(\frac{2}{m})$	29 ($\frac{a+b}{2}$/a:b)·m:2·m
48	c2/a2/m2/m	48	$D_{2h}\bar{c}\alpha\mu\mu$	V_h^{12}	aC2mm	48 $C\frac{2}{mm}(\frac{2}{a})$	48 ($\frac{a+b}{2}$/a:b)·$\tilde{a}$:2·m
SQUARE							
49	p4	58	$C_4\bar{p}$	C_4^1	1P4	54 P(4)11	50 (a:a):4
50	p$\bar{4}$	57	$S_4\bar{p}$	S_4^1	1P$\bar{4}$	49 P($\bar{4}$)11	49 (a:a):$\bar{4}$
51	p4/m	61	$C_{4h}\bar{p}\mu$	C_{4h}^1	mP4	55 P($\frac{4}{m}$)11	51 (a:a):4:m
52	p4/n	62	$D_{4h}\bar{p}\nu$	C_{4h}^2	nP4	56 P($\frac{4}{n}$)11	57 (a:a):4:$\widehat{ab}$
53	p422	67	$D_4\bar{p}11$	D_4^1	1P422	59 P(4)22	55 (a:a):4:2
54	p42$_1$2	68	$D_4\bar{p}21$	D_4^2	1P42$_1$2	60 P(4)2$_1$2	56 (a:a):4:2$_1$
55	p4mm	59	$C_{4v}\bar{p}\mu\mu$	C_{4v}^1	1P4mm	57 P(4)mm	52 (a:a):4·m
56	p4bm	60	$C_{4v}\bar{p}\beta\mu$	C_{4v}^2	1P4gm	58 P(4)bm	59 (a:a):4⊙b
57	p$\bar{4}$2m	63	$D_{2d}\bar{p}\mu1$	V_d^1	1P$\bar{4}$2m	50 P($\bar{4}$)2m	54 (a:a):$\bar{4}$:2
58	p$\bar{4}$2$_1$m	64	$D_{2d}\bar{p}\mu2$	V_d^2	1P$\bar{4}$2$_1$m	51 P($\bar{4}$)2$_1$m	60 (a:a):$\bar{4}$:⊙2$_1$
59	p$\bar{4}$m2	65	$D_{2d}\bar{c}\mu1$	V_d^3	1P$\bar{4}$m2	52 P($\bar{4}$)m2	61 (a:a):$\bar{4}$·m
60	p$\bar{4}$b2	66	$D_{2d}\bar{c}\beta1$	V_d^4	1P$\bar{4}$g2	53 P($\bar{4}$)b2	64 (a:a):$\bar{4}$⊙$\bar{b}$
61	P4/m2/m2/m	69	$D_{4h}\bar{p}\mu\mu\mu$	D_{4h}^1	mP4mm	61 P($\frac{4}{m}$)$\frac{2}{m}\frac{2}{m}$	53 (a:a)·m:4·m
62	p4/n2/b2/m	70	$D_{4h}\bar{p}\nu\beta\mu$	D_{4h}^2	nP4gm	62 P($\frac{4}{n}$)$\frac{2}{b}\frac{2}{m}$	62 (a:a):$\widehat{ab}$:4⊙b
63	p4/m2$_1$/b2/m	71	$D_{4h}\bar{p}\mu\beta\mu$	D_{4h}^3	mP4gm	63 P($\frac{4}{m}$)$\frac{2_1}{b}\frac{2}{m}$	58 (a:a)·m:4⊙b
64	p4/n2$_1$/m2/m	72	$D_{4h}\bar{p}\nu\mu\mu$	D_{4h}^4	nP4mm	64 P($\frac{4}{n}$)$\frac{2_1}{m}\frac{2}{m}$	63 (a:a)·$\widehat{ab}$:4·m
HEXAGONAL							
65	p3	49	$C_3\bar{c}$	C_3^1	1P3	65 P(3)11	65 (a/a):3
66	p$\bar{3}$	50	$S_6\bar{p}$	C_{3i}^1	1P$\bar{3}$	66 P($\bar{3}$)11	67 (a/a):$\bar{3}$
67	p312	54	$D_3\bar{c}1$	D_3^1	1P312	70 P(3)12	72 (a/a):2:3
68	p321	53	$D_3\bar{h}1$	D_3^2	1P321	69 P(3)21	73 (a/a)·2:3
69	p3m1	51	$C_{3v}\bar{c}\mu$	C_{3v}^2	1P3m1	67 P(3)m1	68 (a/a):m·3
70	p31m	52	$C_{3v}\bar{h}\mu$	C_{3v}^1	1P31m	68 P(3)1m	70 (a/a)·m·3
71	p$\bar{3}$12/m	55	$D_{3d}\bar{c}\mu1$	D_{3d}^2	1P$\bar{3}$1m	72 P($\bar{3}$)1m	74 (a/a)·m·$\bar{6}$
72	p$\bar{3}$2/m1	56	$D_{3d}\bar{h}\mu1$	D_{3d}^1	1P$\bar{3}$m1	71 P($\bar{3}$)m1	75 (a/a):m·$\bar{6}$
73	p6	76	$C_6\bar{c}$	C_6^1	1P6	76 P(6)11	76 (a/a):6
74	p$\bar{6}$	73	$C_{3h}\bar{c}\mu$	C_{3h}^1	mP3	73 P($\bar{6}$)11	66 (a/a):3:m

(continued)

TABLE 18
(continued)

	WOOD	WEBER	HERMANN	ALEXANDER AND HERRMANN	NIGGLI	BOHM AND DORNBERGER-SCHIFF	SHUBNIKOV AND KOPTSIK
75	p6/m	78	$C_{6h}\bar{c}\mu$	C_{6h}^1	mP6	77 $P(\frac{6}{m})11$	77 $(a/a)\cdot m{:}6$
76	p622	79	$D_6\bar{c}11$	D_6^1	1P622	79 $P(6)22$	80 $(a/a)\cdot2{:}6$
77	p6mm	77	$C_{6v}\bar{c}\mu\mu$	C_{6v}^1	1P6mm	78 $P(6)mm$	78 $(a/a){:}m\cdot6$
78	p$\bar{6}$m2	74	$D_{3h}\bar{c}\mu\mu$	D_{3h}^1	mP3m2	74 $P(\bar{6})m2$	69 $(a/a){:}m\cdot3{:}m$
79	p$\bar{6}$2m	75	$D_{3h}\bar{h}\mu\mu$	D_{3h}^2	mP32m	75 $P(\bar{6})2m$	71 $(a/a)\cdot m{:}3\cdot m$
80	p6/m2/m2/m	80	$D_{6h}\bar{c}\mu\mu\mu$	D_{6h}^1	mP6mm	80 $P(\frac{6}{m})\frac{2}{m}\frac{2}{m}$	79 $(a/a)\cdot m{:}6\cdot m$

by Wood,[26,27] which will be used in this work. Alongside we give the corresponding numbering and/or symbols introduced by Weber,[15] Hermann,[11,12] Alexander and Herrmann,[13,14] Niggli,[26] Bohm and Dornberger-Schiff,[28] and Shubnikov and Koptsik.[29] For additional symbols and numbering see also Dornberger-Schiff,[30] Cochran,[31] Belov and Tarkhova,[32] Holser,[19] Belov,[33] Chapuis,[34] and Kohler.[35]

The coset representations of the decomposition of each layer group with respect to its subgroup of translations are given in Table 19. The nonprimitive translations are given with respect to the primitive translations t_1 and t_2 given in Figure 5, that is, a nonprimitive translation $at_1 + bt_2$ is denoted by a,b. In the case of layer groups with rectangular centered translation subgroups, the nonzero nonprimitive translations are also given, in parentheses, with respect to the translations

TABLE 19

		1 C_1	4 C_{2z}	25 I	28 m_z
1	p1	0,0			
2	p$\bar{1}$	0,0		0,0	
3	p211	0,0	0,0		
4	pm11	0,0			0,0
5	pb11	0,0			0,½
6	p2/m11	0,0	0,0	0,0	0,0
7	p2/b11	0,0	0,½	0,0	0,½

(continued)

TABLE 19
(continued)

			1 C_1	3 C_{2y}	24 I	27 m_y
	8	p112	0,0	0,0		
	9	p112$_1$	0,0	0,½		
	10	c112	0,0	0,0		
	11	p11m	0,0			0,0
	12	p11a	0,0			½,0
	13	c11m	0,0			0,0
	14	p112/m	0,0	0,0	0,0	0,0
	15	p112$_1$/m	0,0	0,½	0,0	0,½
	16	c112/m	0,0	0,0	0,0	0,0
	17	p112/a	0,0	½,0	0,0	½,0
	18	c112$_1$/a	0,0	½,½	0,0	½,½

		1 C_1	2 C_{2x}	3 C_{2y}	4 C_{2z}
19	p222	0,0	0,0	0,0	0,0
20	p222$_1$	0,0	0,½	0,½	0,0
21	p22$_1$2$_1$	0,0	½,½	½,½	0,0
22	c222	0,0	0,0	0,0	0,0

		1 C_1	2 C_{2x}	3 C_{2y}	4 C_{2z}	25 I	26 m_x	27 m_y	28 m_z
23	p2mm	0,0			0,0		0,0	0,0	
24	pmm2	0,0		0,0			0,0		0,0
25	pm2$_1$a	0,0	½,0					½,0	0,0
26	pbm2$_1$	0,0		0,½			0,0		0,½
27	pbb2	0,0		0,0			0,½		0,½
28	p2ma	0,0			0,0		½,0	½,0	
29	pam2	0,0		0,0			½,0		½,0
30	pab2$_1$	0,0		0,½			½,½		½,0
31	pnb2	0,0		0,0			½,½		½,½
32	pnm2$_1$	0,0		½,½			0,0		½,½
33	p2ba	0,0			0,0		½,½	½,½	
34	c2mm	0,0			0,0		0,0	0,0	
35	cmm2	0,0		0,0			0,0		0,0
36	cam2	0,0		0,0			−½,½ (0,½)		−½,½ (0,½)

(continued)

TABLE 19
(continued)

		1 C_1	2 C_{2x}	3 C_{2y}	4 C_{2z}	25 I	26 m_x	27 m_y	28 m_z
37	p2/m2/m2/m	0,0	0,0	0,0	0,0	0,0	0,0	0,0	0,0
38	p2/a2/m2/a	0,0	0,0	½,0	½,0	0,0	0,0	½,0	½,0
39	p2/n2/b2/a	0,0	0,0	0,0	0,0	½,½	½,½	½,½	½,½
40	$p2/m2_1/m2/a$	0,0	½,0	½,0	0,0	0,0	½,0	½,0	0,0
41	$p2/a2_1/m2/m$	0,0	½,0	0,0	½,0	0,0	½,0	0,0	½,0
42	$p2/n2/m2_1/a$	0,0	0,0	½,½	½,½	0,0	0,0	½,½	½,½
43	$p2/a2/b2_1/a$	0,0	0,½	½,½	½,0	0,0	0,½	½,½	½,0
44	$p2/m2_1/b2_1/a$	0,0	½,½	½,½	0,0	0,0	½,½	½,½	0,0
45	$p2/a2_1/b2_1/m$	0,0	½,½	0,½	½,0	0,0	½,½	0,½	½,0
46	$p2/n2_1/m2_1/m$	0,0	½,½	½,½	0,0	½,½	0,0	0,0	½,½
47	c2/m2/m2/m	0,0	0,0	0,0	0,0	0,0	0,0	0,0	0,0
48	c2/a2/m2/m	0,0	0,0	½,½ (½,0)	½,½ (½,0)	0,0	0,0	½,½ (½,0)	½,½ (½,0)

		1 C_1	14 C_{4z}	4 C_{2z}	15 C_{4z}^{-1}	25 I	38 S_{4z}^{-1}	28 m_z	39 S_{4z}
49	p4	0,0	0,0	0,0	0,0				
50	$p\bar{4}$	0,0		0,0			0,0		0,0
51	p4/m	0,0	0,0	0,0	0,0	0,0	0,0	0,0	0,0
52	p4/n	0,0	½,½	0,0	½,½	½,½	0,0	½,½	0,0

		1 C_1	14 C_{4z}	4 C_{2z}	15 C_{4z}^{-1}	2 C_{2x}	3 C_{2y}	16 C_{2xy}	13 $C_{2\bar{x}y}$	25 I	38 S_{4z}^{-1}	28 m_z	39 S_{4z}	26 m_x	27 m_y	40 m_{xy}	37 $m_{\bar{x}y}$
53	p422	0,0	0,0	0,0	0,0	0,0	0,0	0,0	0,0								
54	$p42_12$	0,0	½,½	0,0	½,½	0,0	0,0	½,½	½,½								
55	p4mm	0,0	0,0	0,0	0,0									0,0	0,0	0,0	0,0
56	p4bm	0,0	0,0	0,0	0,0									½,½	½,½	½,½	½,½
57	$p\bar{4}2m$	0,0		0,0		0,0	0,0				0,0		0,0			0,0	0,0
58	$p\bar{4}2_1m$	0,0		0,0		½,½	½,½				0,0		0,0			½,½	½,½
59	$p\bar{4}m2$	0,0		0,0				0,0	0,0		0,0		0,0	0,0	0,0		
60	$p\bar{4}b2$	0,0		0,0				½,½	½,½		0,0		0,0	½,½	½,½		
61	p4/m2/m2/m	0,0	0,0	0,0	0,0	0,0	0,0	0,0	0,0	0,0	0,0	0,0	0,0	0,0	0,0	0,0	0,0
62	p4/n2/b2/m	0,0	0,0	0,0	0,0	0,0	0,0	0,0	0,0	½,½	½,½	½,½	½,½	½,½	½,½	½,½	½,½
63	$p4/m2_1/b2/m$	0,0	0,0	0,0	0,0	½,½	½,½	½,½	½,½	0,0	0,0	0,0	0,0	½,½	½,½	½,½	½,½
64	$p4/n2_1/m2/m$	0,0	½,½	0,0	½,½	½,½	½,½	0,0	0,0	½,½	0,0	½,½	0,0	0,0	0,0	½,½	½,½

(continued)

TABLE 19
(continued)

		1 c_1	3 c_3	5 c_3^{-1}	7 c_{2x}	9 c_{2xy}	11 c_{2y}	8 c_{21}	10 c_{22}	12 c_{23}	13 I	15 s_6^{-1}	17 s_6	19 m_x	21 m_{xy}	23 m_y	20 m_1	22 m_2	24 m_3
65	p3	0,0	0,0	0,0															
66	p$\bar{3}$	0,0	0,0	0,0							0,0	0,0	0,0						
67	p312	0,0	0,0	0,0	0,0	0,0	0,0												
68	p321	0,0	0,0	0,0	0,0	0,0	0,0	0,0											
69	p3m1	0,0	0,0	0,0										0,0	0,0	0,0			
70	p31m	0,0	0,0	0,0													0,0	0,0	0,0
71	p$\bar{3}$12/m	0,0	0,0	0,0				0,0	0,0	0,0	0,0	0,0	0,0				0,0	0,0	0,0
72	p$\bar{3}$2/m1	0,0	0,0	0,0	0,0	0,0	0,0	0,0			0,0	0,0	0,0	0,0	0,0	0,0			

		1 c_1	2 c_6	3 c_3	4 c_2	5 c_3^{-1}	6 c_6^{-1}	7 c_{2x}	9 c_{2xy}	11 c_{2y}	8 c_{21}	10 c_{22}	12 c_{23}
73	p6	0,0	0,0	0,0	0,0	0,0	0,0						
74	p$\bar{6}$	0,0		0,0		0,0							
75	p6/m	0,0	0,0	0,0	0,0	0,0	0,0						
76	p622	0,0	0,0	0,0	0,0	0,0	0,0	0,0	0,0	0,0	0,0	0,0	0,0
77	p6mm	0,0	0,0	0,0	0,0	0,0	0,0						
78	p$\bar{6}$m2	0,0		0,0		0,0					0,0	0,0	0,0
79	p$\bar{6}$2m	0,0		0,0		0,0		0,0	0,0	0,0			
80	p6/m2/m2/m	0,0	0,0	0,0	0,0	0,0	0,0	0,0	0,0	0,0	0,0	0,0	0,0

		13 I	14 s_3^{-1}	15 s_6^{-1}	16 m_z	17 s_6	18 s_3	19 m_x	21 m_{xy}	23 m_y	20 m_1	22 m_2	24 m_3
73	p6												
74	p$\bar{6}$		0,0		0,0		0,0						
75	p6/m	0,0	0,0	0,0	0,0	0,0	0,0						
76	p622												
77	p6mm							0,0	0,0	0,0	0,0	0,0	0,0
78	p$\bar{6}$m2		0,0		0,0		0,0	0,0	0,0	0,0			
79	p$\bar{6}$2m		0,0		0,0		0,0				0,0	0,0	0,0
80	p6/m2/m2/m	0,0	0,0	0,0	0,0	0,0	0,0	0,0	0,0	0,0	0,0	0,0	0,0

$t_1 + t_2$ and $t_1 - t_2$. These are the generators of the so-called conventional rectangular translation subgroup of the rectangular centered translation group.

Each of the eighty layer groups is related to a three-dimensional space group. Let G denote a three-dimensional space group and T(1) a one-dimensional translation subgroup of G. Each layer group is isomorphic to a factor group G/T(1). This relationship has been considered in detail by Wood.[26],[27] In Table 20, using the numbering of Table 18 and Reference 1, we list each of the eighty layer groups along with the three-dimensional space group to which it is related.

TABLE 20

Layer Group	Space Group	Layer Group	Space Group
1	1	41	51
2	2	42	53
3	3	43	54
4	6	44	55
5	7	45	57
6	10	46	59
7	13	47	65
8	3	48	67
9	4	49	75
10	5	50	81
11	6	51	83
12	7	52	85
13	8	53	89
14	10	54	90
15	11	55	99
16	12	56	100
17	13	57	111
18	14	58	113
19	16	59	115
20	17	60	117
21	18	61	123
22	21	62	125
23	25	63	127
24	25	64	129
25	26	65	143
26	26	66	147
27	27	67	149
28	28	68	150
29	28	69	156
30	29	70	157
31	30	71	162
32	31	72	164
33	32	73	168
34	35	74	174
35	38	75	175
36	39	76	177
37	47	77	183
38	49	78	187
39	50	79	189
40	51	80	191

5. k Labeling of Irreducible Representations

For each translation group generated by t_1 and t_2 there exists a "reciprocal lattice" generated by the vectors K_1 and K_2 which satisfies the relationship $K_i \cdot t_j = 2\pi\delta_{ij}$, $i,j = 1,2$. In Table 21 we list, for each of the five classes of two-dimensional translation groups, the generating translations t_1, t_2, K_1, and K_2. The components of these translations are given in an orthogonal coordinate system, except in the hexagonal case, where they are given in a hexagonal coordinate system. In Figure 6 we give the generators of the reciprocal lattice and the corresponding Brillouin zone for each of the five classes of two-dimensional translation groups. The action of each element of the holosymmetric point group on the generators of the reciprocal lattice is given in Table 22.

In each Brillouin zone there is a basic domain Ω such that the whole Brillouin zone is generated from Ω by the elements of the holosymmetric point group of the corresponding translation group. For each layer group there is a representation domain such that the whole Brillouin zone is generated from the representation domain by the elements of the isogonal point group of that layer group.[9] The isogonal point group is the point group consisting of all the proper and improper rotations of the layer group. The representation domain of a layer group is a small integral multiple of the basic domain.

TABLE 21

Oblique	p	$t_1 = (a,-c)$	$K_1 = 2\pi\,(\frac{1}{a},\,0)$
		$t_2 = (0,b)$	$K_2 = 2\pi\,(\frac{c}{ab},\,\frac{1}{b})$
Rectangular	p	$t_1 = (a,0)$	$K_1 = 2\pi\,(\frac{1}{a},0)$
		$t_2 = (0,b)$	$K_2 = 2\pi\,(0,\frac{1}{b})$
Rectangular	c	$t_1 = (\frac{a}{2},-\frac{b}{2})$	$K_1 = 2\pi\,(\frac{1}{a},-\frac{1}{b})$
		$t_2 = (\frac{a}{2},\frac{b}{2})$	$K_2 = 2\pi\,(\frac{1}{a},\frac{1}{b})$
Square	p	$t_1 = (a,0)$	$K_1 = 2\pi\,(\frac{1}{a},\,0)$
		$t_2 = (0,a)$	$K_2 = 2\pi\,(0,\frac{1}{a})$
Hexagonal	p	$t_1 = (a,0)$	$K_1 = 2\pi\,(\frac{4}{3a},\frac{2}{3a})$
		$t_2 = (0,a)$	$K_2 = 2\pi\,(\frac{2}{3a},\,\frac{4}{3a})$

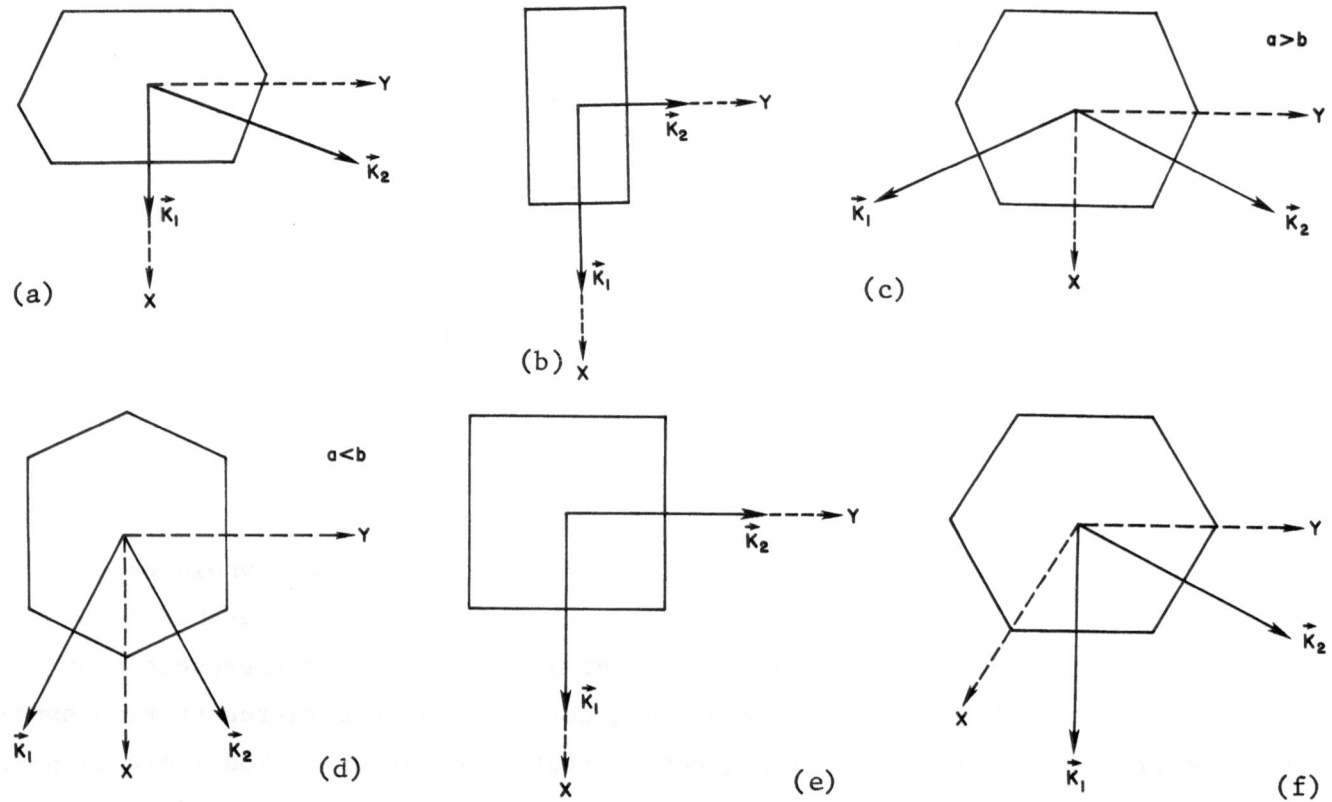

Figure 6. Generators of the two-dimensional reciprocal lattices:
(a) Oblique p, (b) Rectangular p, (c,d) Rectangular c,
(e) Square p, and (f) Hexagonal p.

TABLE 22

Oblique p			
C_1	1	K_1	K_2
C_{2z}	4	$-K_1$	$-K_2$
I	25	$-K_1$	$-K_2$
m_z	28	K_1	K_2

Rectangular p			
C_1	1	K_1	K_2
C_{2x}	2	K_1	$-K_2$
C_{2y}	3	$-K_1$	K_2
C_{2z}	4	$-K_1$	$-K_2$
I	25	$-K_1$	$-K_2$
m_x	26	$-K_1$	K_2
m_y	27	K_1	$-K_2$
m_z	28	K_1	K_2

Rectangular c			
C_1	1	K_1	K_2
C_{2x}	2	K_2	K_1
C_{2y}	3	$-K_2$	$-K_1$
C_{2z}	4	$-K_1$	$-K_2$
I	25	$-K_1$	$-K_2$
m_x	26	$-K_2$	$-K_1$
m_y	27	K_2	K_1
m_z	28	K_1	K_2

(continued)

TABLE 22
(continued)

Square p

C_1	1	K_1	K_2
C_{2x}	2	K_1	$-K_2$
C_{2y}	3	$-K_1$	K_2
C_{2z}	4	$-K_1$	$-K_2$
$C_{2\bar{x}y}$	13	$-K_2$	$-K_1$
C_{4z}	14	K_2	$-K_1$
C_{4z}^{-1}	15	$-K_2$	K_1
C_{2xy}	16	K_2	K_1
I	25	$-K_1$	$-K_2$
m_x	26	$-K_1$	K_2
m_y	27	K_1	$-K_2$
m_z	28	K_1	K_2
$m_{\bar{x}y}$	37	K_2	K_1
S_{4z}^{-1}	38	$-K_2$	K_1
S_{4z}	39	K_2	$-K_1$
m_{xy}	40	$-K_2$	$-K_1$

Hexagonal p

C_1	1	K_1	K_2	I	13	$-K_1$	$-K_2$
C_6	2	K_2	$-K_1+K_2$	S_3^{-1}	14	$-K_2$	K_1-K_2
C_3	3	$-K_1+K_2$	$-K_1$	S_6^{-1}	15	K_1-K_2	K_1
C_2	4	$-K_1$	$-K_2$	m_z	16	K_1	K_2
C_3^{-1}	5	$-K_2$	K_1-K_2	S_6	17	K_2	$-K_1+K_2$
C_6^{-1}	6	K_1-K_2	K_1	S_3	18	$-K_1+K_2$	$-K_1$
C_{2x}	7	K_1-K_2	$-K_2$	m_x	19	$-K_1+K_2$	K_2
C_{21}	8	K_1	K_1-K_2	m_1	20	$-K_1$	$-K_1+K_2$
C_{2xy}	9	K_2	K_1	m_{xy}	21	$-K_2$	$-K_1$
C_{22}	10	$-K_1+K_2$	K_2	m_2	22	K_1-K_2	$-K_2$
C_{2y}	11	$-K_1$	$-K_1+K_2$	m_y	23	K_1	K_1-K_2
C_{23}	12	$-K_2$	$-K_1$	m_3	24	K_2	K_1

TABLE 23

Layer Group		R_i			
Oblique p					
1,4,5	$C_1(1)$	$I(25)$			
2,3,6,7	$C_1(1)$				
Rectangular p					
8,9,24,26,27,29,30,31,32	$C_1(1)$	$m_y(27)$			
11,12,25	$C_1(1)$	$m_x(26)$			
14,15,17,18,19,20,21 23,28,33,37,38,39,40 41,42,43,44,45,46	$C_1(1)$				
Rectangular c					
10,35,36	$C_1(1)$	$m_y(27)$			
13	$C_1(1)$	$m_x(26)$			
16,22,34,47,48	$C_1(1)$				
Square p					
49,50,51,52	$C_1(1)$	$m_{\overline{xy}}(37)$			
53,54,55,56,57,58,59,60 61,62,63,64	$C_1(1)$				
Hexagonal p					
65,74	$C_1(1)$	$m_y(23)$	$m_3(24)$	$C_6(2)$	
66,68,70,73,75,79	$C_1(1)$	$m_y(23)$			
67,69,78	$C_1(1)$	$m_3(24)$			
71,72,76,77,80	$C_1(1)$				

In Table 23 we list for each of the eighty layer groups the elements R_i of the isogonal point group which generate the representation domain from the basic domain. The areas of the Brillouin zone generated by elements R_i from the basic domain Ω are denoted by $R_i\Omega$ in Figure 7.

The labeling of irreducible representations of the layer groups is based on the use of the wave vector k.[36] The representation domain is the smallest fraction of the Brillouin zone which contains at least one k vector from the label of all irreducible representations of a layer group.

In Table 24 we tabulate the symbols and coordinates of all wave vectors k used in labeling the irreducible representations of the layer groups. These wave vectors

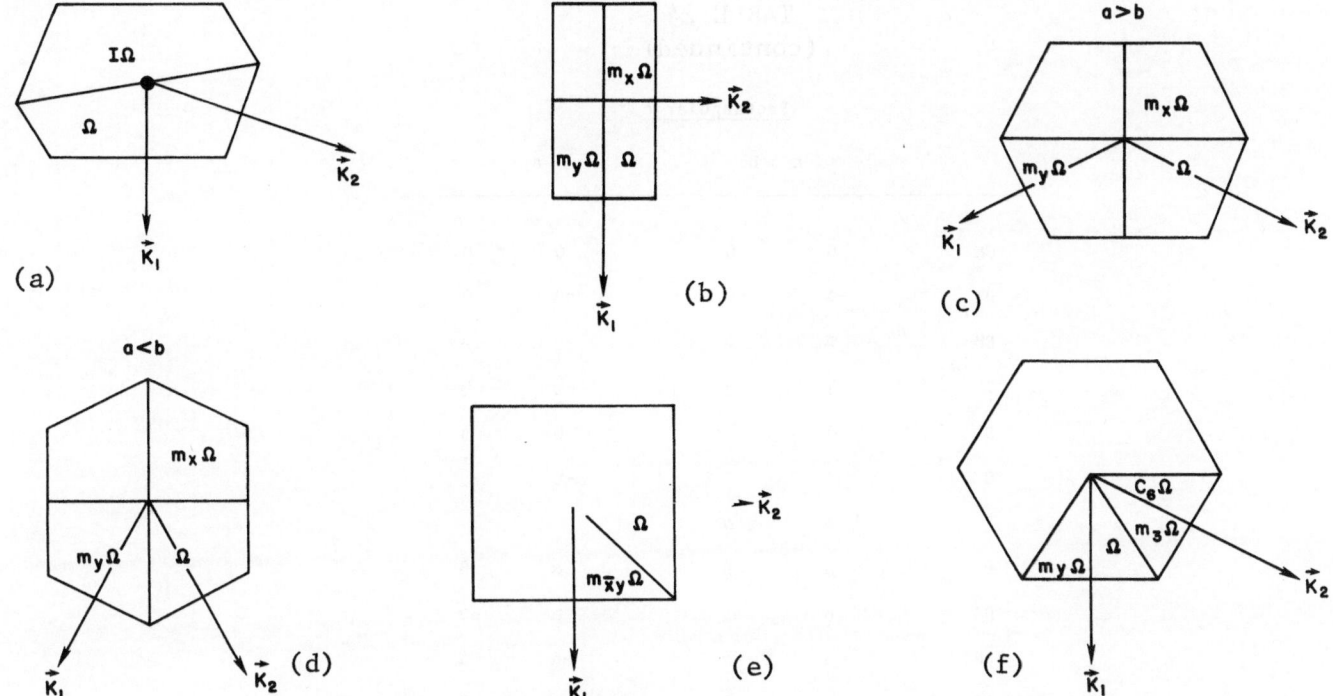

Figure 7.   Brillouin zone, basic domain, and representation domain:
(a) Oblique p, (b) Rectangular p, (c,d) Rectangular c,
(e) Square p, and (f) Hexagonal p.

TABLE 24

Oblique p				Rectangular p		
GM	0	0		GM	0	0
A	½	-½		DT	0	α
B	½	0		SM	α	0
Y	0	½		X	½	0
F	α	β		Y	0	½
FA	-α	-β		S	½	½
				D	½	α
				C	α	½
				V	α	β
				TA	0	-α
				SN	-α	0
				DA	½	-α
				CA	-α	½
				VA	α	-β
				VB	-α	β

(continued)

TABLE 24
(continued)

Rectangular c

	a > b		a < b	
GM	0	0	0	0
DT	$-\alpha$	α	$-\alpha$	α
SM	α	α	α	α
Y	$\frac{1}{2}$	$\frac{1}{2}$	$-\frac{1}{2}$	$\frac{1}{2}$
S	0	$\frac{1}{2}$	0	$\frac{1}{2}$
C			$\alpha-\frac{1}{2}$	$\alpha+\frac{1}{2}$
F	$\frac{1}{2}-\alpha$	$\frac{1}{2}+\alpha$		
P	α	β	α	β
DA	α	$-\alpha$	α	$-\alpha$
SN	$-\alpha$	$-\alpha$	$-\alpha$	$-\alpha$
SA	$-\frac{1}{2}$	0	$-\frac{1}{2}$	0
CA			$-\alpha-\frac{1}{2}$	$-\alpha+\frac{1}{2}$
FA	$\frac{1}{2}+\alpha$	$\frac{1}{2}-\alpha$		
PA	β	α	β	α
PB	$-\beta$	$-\alpha$	$-\beta$	$-\alpha$

Square p

GM	0	0
SM	α	α
DT	0	α
M	$\frac{1}{2}$	$\frac{1}{2}$
X	0	$\frac{1}{2}$
Y	α	$\frac{1}{2}$
C	α	β
DA	β	α

Hexagonal p

GM	0	0
LD	α	α
SM	α	0
K	$\frac{1}{3}$	$\frac{1}{3}$
M	$\frac{1}{2}$	0
T	$\frac{1}{2}-\alpha$	2α
B	α	β
LE	2α	$-\alpha$
SN	0	α
KA	$\frac{2}{3}$	$-\frac{1}{3}$
TA	$\frac{1}{2}+\alpha$	-2α
BA	$\alpha+\beta$	$-\beta$
BB	β	α
BC	$-\beta$	$\alpha+\beta$

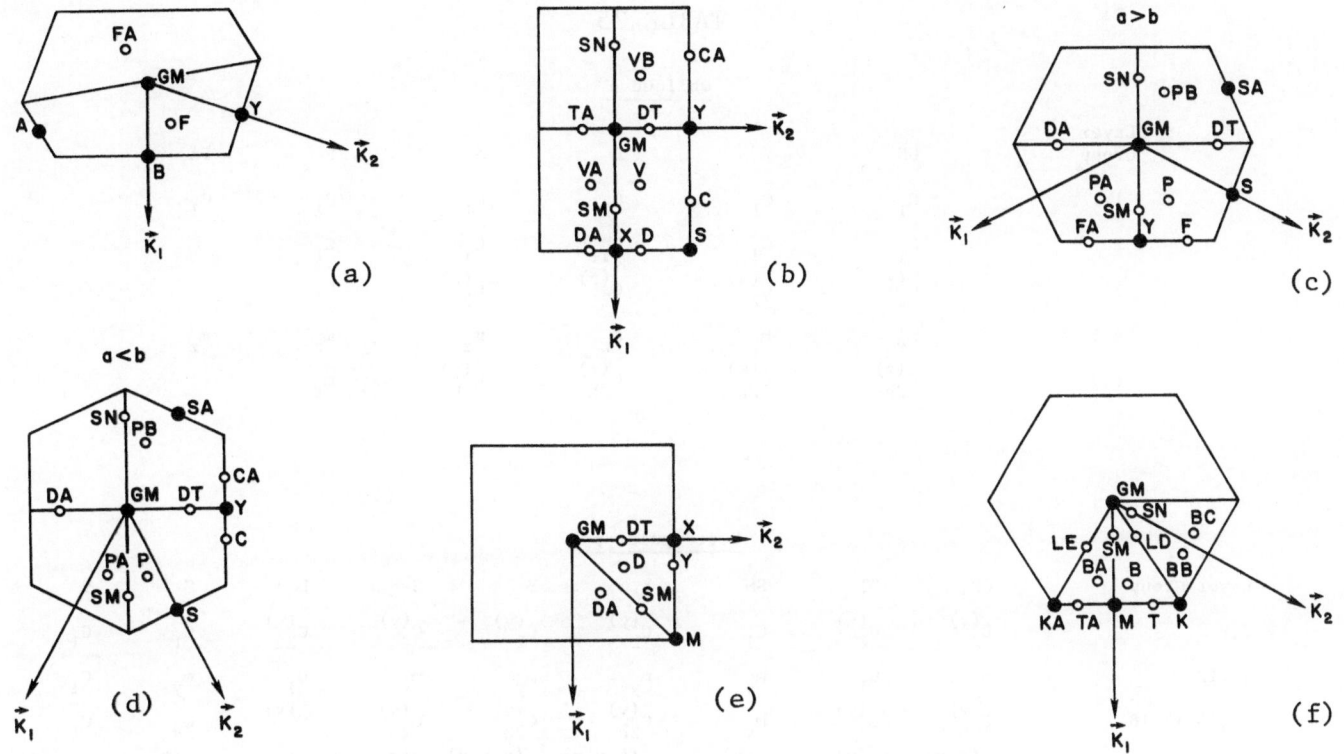

Figure 8. Wave vectors k: (a) Oblique p, (b) Rectangular p,
(c,d) Rectangular c, (e) Square p, and (f) Hexagonal p.

k are shown in Figure 8. The ranges of the variables α and β are such that the position of wave vectors in the basic domain remains in the basic domain.

The wave vectors k used in the labeling of the irreducible representations of each of the eighty layer groups are given in Table 25. The point group of the wave vector k is that subgroup of all elements R of the isogonal point group of the layer group for which Rk is equivalent to k, i.e., for which Rk = k + K, where K is a reciprocal lattice vector. The point group of the wave vector k for each layer group and wave vector k is given in Table 25 at the intersection of the row indexed by the layer group number and column corresponding to the wave vector k.

6. Character Tables

The irreducible representations of a layer group are determined from the irreducible representations of the group G^k of the wave vector k.[8,9,37] Writing G^k as

TABLE 25

Oblique p

Layer Group	GM	Y	B	A	F	FA
1	C_1	C_1	C_1	C_1	C_1	C_1
2	C_i	C_i	C_i	C_i	C_1	
3	$C_2^{(z)}$	$C_2^{(z)}$	$C_2^{(z)}$	$C_2^{(z)}$	C_1	
4,5	m_z	m_z	m_z	m_z	m_z	m_z
6,7	$C_{2h}^{(z)}$	$C_{2h}^{(z)}$	$C_{2h}^{(z)}$	$C_{2h}^{(z)}$	m_z	

Rectangular p

Layer Group	GM	DT	SM	X	Y	S	D	C	V
8,9	$C_2^{(y)}$	$C_2^{(y)}$	C_1	$C_2^{(y)}$	$C_2^{(y)}$	$C_2^{(y)}$	$C_2^{(y)}$	C_1	C_1
11,12	m_y	C_1	m_y	m_y	m_y	m_y	C_1	m_y	C_1
14,15,17,18	$C_{2h}^{(y)}$	$C_2^{(y)}$	m_y	$C_{2h}^{(y)}$	$C_{2h}^{(y)}$	$C_{2h}^{(y)}$	$C_2^{(y)}$	m_y	C_1
19,20,21	$D_2^{(x,y,z)}$	$C_2^{(y)}$	$C_2^{(x)}$	$D_2^{(x,y,z)}$	$D_2^{(x,y,z)}$	$D_2^{(x,y,z)}$	$C_2^{(y)}$	$C_2^{(x)}$	C_1
23,28,33	$C_{2v}^{(z,x,y)}$	m_x	m_y	$C_{2v}^{(z,x,y)}$	$C_{2v}^{(z,x,y)}$	$C_{2v}^{(z,x,y)}$	m_x	m_y	C_1
24,26,27,29,30,31,32	$C_{2v}^{(y)}$	$C_{2v}^{(y)}$	m_z	$C_{2v}^{(y)}$	$C_{2v}^{(y)}$	$C_{2v}^{(y)}$	$C_{2v}^{(y)}$	m_z	m_z
25	$C_{2v}^{(x)}$	m_z	$C_{2v}^{(x)}$	$C_{2v}^{(x)}$	$C_{2v}^{(x)}$	$C_{2v}^{(x)}$	m_z	$C_{2v}^{(x)}$	m_z
37,38,39,40,41 42,43,44,45,46	$D_{2h}^{(x,y,z)}$	$C_{2v}^{(y)}$	$C_{2v}^{(x)}$	$D_{2h}^{(x,y,z)}$	$D_{2h}^{(x,y,z)}$	$D_{2h}^{(x,y,z)}$	$C_{2v}^{(y)}$	$C_{2v}^{(x)}$	m_z

Layer Group	TA	SN	DA	CA	VA	VB
8,9	$C_2^{(y)}$		$C_2^{(y)}$		C_1	
11,12		C_1		m_y		C_1
14,15,17,18						
19,20,21						
23,28,33						
24,26,27,29,30,31 32	$C_{2v}^{(y)}$		$C_{2v}^{(y)}$			m_z
25		$C_{2v}^{(x)}$		$C_{2v}^{(x)}$		m_z
37,38,39,40,41 42,43,44,45,46						

(continued)

TABLE 25
(continued)

Rectangular c

Layer Group	GM	DT	SM	Y	S	C	F	P
10	$C_2^{(y)}$	$C_2^{(y)}$	C_1	$C_2^{(y)}$	$C_2^{(y)}$	C_1	$C_2^{(y)}$	C_1
13	m_y	C_1	m_y	m_y	C_1	m_y	C_1	C_1
16	$C_{2h}^{(y)}$	$C_2^{(y)}$	m_y	$C_{2h}^{(y)}$	C_i	m_y	$C_2^{(y)}$	C_1
22	$D_2^{(x,y,z)}$	$C_2^{(y)}$	$C_2^{(x)}$	$D_2^{(x,y,z)}$	$C_2^{(z)}$	$C_2^{(x)}$	$C_2^{(y)}$	C_1
34	$C_{2v}^{(z,x,y)}$	m_x	m_y	$C_{2v}^{(z,x,y)}$	$C_2^{(z)}$	m_y	m_x	C_1
35,36	$C_{2v}^{(y)}$	$C_{2v}^{(y)}$	m_z	$C_{2v}^{(y)}$	m_z	$C_{2v}^{(y)}$	$C_{2v}^{(y)}$	m_z
47,48	$D_{2h}^{(x,y,z)}$	$C_{2v}^{(y)}$	$C_{2v}^{(x)}$	$D_{2h}^{(x,y,z)}$	$C_{2h}^{(z)}$	$C_{2v}^{(y)}$	$C_{2v}^{(y)}$	m_z

Layer Group	DA	SN	SA	CA	FA	PA	PB
10	$C_2^{(y)}$				$C_2^{(y)}$	C_1	
13		m_y		m_y			C_1
16							
22							
34							
35,36	$C_{2v}^{(y)}$				$C_{2v}^{(y)}$	m_z	
47,48							

Square p

Layer Group	GM	SM	DT	M	X	Y	D	DA
49	C_4	C_1	C_1	C_4	$C_2^{(z)}$	C_1	C_1	C_1
50	S_4	C_1	C_1	S_4	$C_2^{(z)}$	C_1	C_1	C_1
51,52	C_{4h}	m_z	m_z	C_{4h}	$C_{2h}^{(z)}$	m_z	m_z	m_z
53,54	D_4	$C_2^{(xy)}$	$C_2^{(y)}$	D_4	$D_2^{(z,x,y)}$	$C_2^{(x)}$	C_1	
55,56	C_{4v}	$m_{\overline{xy}}$	m_x	C_{4v}	$C_{2v}^{(z,x,y)}$	m_y	C_1	
57,58	$D_{2d}^{(z,x,y)}$	$m_{\overline{xy}}$	$C_2^{(y)}$	$D_{2d}^{(z,x,y)}$	$D_2^{(z,x,y)}$	$C_2^{(x)}$	C_1	
59,60	$D_{2d}^{(z,xy,\overline{xy})}$	$C_2^{(xy)}$	m_x	$D_{2d}^{(z,xy,xy)}$	$C_{2v}^{(z,x,y)}$	m_y	C_1	
61,62,63,64	D_{4h}	$C_{2v}^{(xy)}$	$C_{2v}^{(y)}$	D_{4h}	$D_{2h}^{(z,x,y)}$	$C_{2v}^{(x)}$	m_z	

TABLE 25
(continued)

Hexagonal p

Layer Group	GM	LD	SM	K	M	T	B
65	C_3	C_1	C_1	C_3	C_1	C_1	C_1
66	C_{3i}	C_1	C_1	C_3	C_i	C_1	C_1
67	$D_3^{(1,2,3)}$	C_1	$C_2^{(1)}$	C_3	$C_2^{(1)}$	C_1	C_1
68	$D_3^{(x,xy,y)}$	$C_2^{(xy)}$	C_1	$D_3^{(x,xy,y)}$	$C_2^{(y)}$	$C_2^{(y)}$	C_1
69	$C_{3v}^{(x,xy,y)}$	C_1	m_y	C_3	m_y	C_1	C_1
70	$C_{3v}^{(1,2,3)}$	m_3	C_1	$C_{3v}^{(1,2,3)}$	m_1	m_1	C_1
71	$D_{3d}^{(1,2,3)}$	m_3	$C_2^{(1)}$	$C_{3v}^{(1,2,3)}$	$C_{2h}^{(1)}$	m_1	C_1
72	$D_{3d}^{(x,xy,y)}$	$C_2^{(xy)}$	m_y	$D_3^{(x,xy,y)}$	$C_{2h}^{(y)}$	$C_2^{(y)}$	C_1
73	C_6	C_1	C_1	C_3	$C_2^{(z)}$	C_1	C_1
74	C_{3h}	m_z	m_z	C_{3h}	m_z	m_z	m_z
75	C_{6h}	m_z	m_z	C_{3h}	$C_{2h}^{(z)}$	m_z	m_z
76	D_6	$C_2^{(xy)}$	$C_2^{(1)}$	$D_3^{(x,xy,y)}$	$D_2^{(y,1,z)}$	$C_2^{(y)}$	C_1
77	C_{6v}	m_3	m_y	$C_{3v}^{(1,2,3)}$	$C_{2v}^{(z,y,1)}$	m_1	C_1
78	$D_{3h}^{(1,2,3)}$	m_z	$C_{2v}^{(1)}$	C_{3h}	$C_{2v}^{(1)}$	m_z	m_z
79	$D_{3h}^{(x,xy,y)}$	$C_{2v}^{(xy)}$	m_z	$D_{3h}^{(x,xy,y)}$	$C_{2v}^{(y)}$	$C_{2v}^{(y)}$	m_z
80	D_{6h}	$C_{2v}^{(xy)}$	$C_{2v}^{(1)}$	$D_{3h}^{(x,xy,y)}$	$D_{2h}^{(z,y,1)}$	$C_{2v}^{(y)}$	m_z

Layer Group	LE	SN	KA	TA	BA	BB	BC
65	C_1	C_1	C_3	C_1	C_1	C_1	C_1
66					C_1		
67		$C_2^{(2)}$				C_1	
68	$C_2^{(x)}$		$D_3^{(x,xy,y)}$	$C_2^{(y)}$	C_1		
69		m_x				C_1	
70	m_y		$C_{3v}^{(1,2,3)}$	m_1	C_1		
71							
72							
73					C_1		
74	m_z	m_z	C_{3h}	m_z	m_z	m_z	m_z
75					m_z		
76							
77							
78		$C_{2v}^{(2)}$				m_z	
79	$C_{2v}^{(x)}$		$D_{3h}^{(x,xy,y)}$	$C_{2v}^{(y)}$	m_z		
80							

a coset decomposition with respect to the translational subgroup T, we have

$$G^k = T + (\beta_2|\tau_2)T + (\beta_3|\tau_3)T + \cdots + (\beta_n|\tau_n)T$$

where the proper and improper rotations β_i, $i = 1,2,\ldots,n$, $\beta_1 = C_1$, are the elements of the point group of the wave vector k (see Table 25) and τ_i, $i = 1,2,\ldots,n$ are the nonprimitive translations associated with β_i (see Table 19).

The irreducible representations D_j^k of the group G^k of the wave vector k are such that

$$D_j^k(\beta_i|\tau_i + t) = D_j^k(\beta_i|\tau_i)e^{ik\cdot t}$$

where t is a translation contained in T. We have tabulated in Table 26 (see Appendix I), for each layer group, the characters $\chi_j^k(\beta_i|\tau_i)$ for all elements $(\beta_i|\tau_i)$ of each irreducible representation D_j^k for each wave vector k listed in Table 24. The rotational and translational parts of elements $(\beta_i|\tau_i)$ are given numerical symbols in Table 25. The rotational parts use the numbering of Tables 2 and 3. This is summarized in Table 27 along with the numerical symbols of the nonprimitive translation τ_i of nonsymmorphic layer groups. In Table 26 elements of the form $(\beta_i|00)$ are denoted by the numerical symbol of β_i, elements $(\beta_i|\tau_i)$ by the numerical symbol of β_i followed by a comma and the numerical symbol for τ_i. The symbols used in Table 26 are $J = \sqrt{-1}$, $A = \sqrt{2}$, $B = \sqrt{3}$, $U = e^{i\pi/6}$, $V = e^{i\pi/4}$, $W = e^{i2\pi/3}$. An asterisk denotes complex conjugation, and $T = e^{ik\cdot\tau}$, where τ is a nonprimitive translation. The characters are obtained by utilizing the isomorphism of each layer group with a factor group G/T(I) of a three-dimensional space group G and referring to tables of the characters of the space groups.[9]

For example, for group 21, $p22_12_1$, and k vector SM, from Table 25 we have the following character table:

SM		1	2	3	4
	1	1	1	1	1
	49	1	1	-1	-1
	2,1	1,T	-1,T	-1,TJ	1,TJ
	50,1	1,T	-1,T	1,TJ	-1,TJ

TABLE 27

Rotational Elements of Layer Groups 1-64

R	R	$\overline{E}R$	RI	RI	$\overline{E}RI$
C_1	1	49	I	25	73
C_{2x}	2	50	m_x	26	74
C_{2y}	3	51	m_y	27	75
C_{2z}	4	52	m_z	28	76
$C_{2\overline{xy}}$	13	61	$m_{\overline{xy}}$	37	85
C_{4z}	14	62	S_{4z}^{-1}	38	86
C_{4z}^{-1}	15	63	S_{4z}	39	87
C_{2xy}	16	64	m_{xy}	40	88

Rotational Elements of Layer Groups 65-80

R	R	$\overline{E}R$	RI	RI	$\overline{E}RI$
C_1	1	49	I	13	61
C_6	2	50	S_3^{-1}	14	62
C_3	3	51	S_6^{-1}	15	63
C_2	4	52	m_z	16	64
C_3^{-1}	5	53	S_6	17	65
C_6^{-1}	6	54	S_3	18	66
C_{2x}	7	55	m_x	19	67
C_{21}	8	56	m_1	20	68
C_{2xy}	9	57	m_{xy}	21	69
C_{22}	10	58	m_2	22	70
C_{2y}	11	59	m_y	23	71
C_{23}	12	60	m_3	24	72

Nonprimitive Translations of Nonsymmorphic Layer Groups

Layer Group	1	2	3
5,7,9,15,20,26,27	(0,1/2)		
12,17,25,28,29,38,40,41	(1/2,0)		
18,21,31,32,33,39, 42,44,46,48,52,54, 56,58,60,62,63,64	(1/2,1/2)		
36	(-1/2,1/2)		
30,43	(0,1/2)	(1/2,1/2)	(1/2,0)
45	(0,1/2)	(1/2,0)	(1/2,1/2)

The coordinates of the k vector SM are $(\alpha,0)$ for this rectangular layer group and are found in Table 24. The point group of the wave vector k is given in Table 25. The elements of the point group of the wave vector k and their corresponding non-primitive translations are listed in the left-hand column in numerical form, and are found in Table 27. For the above we have $1 = (C_1|00)$, $2,1 = (\bar{C}_{2x}|\frac{1}{2}\frac{1}{2})$, $49 = (\bar{C}_1|00)$, and $50,1 = (\bar{C}_{2x}|\frac{1}{2}\frac{1}{2})$. The four irreducible representations are indexed by the number in the top row. The character $\chi_j^{SM}(\beta_i|\tau_i)$ is given at the intersection of the row indexed by the numerical equivalent of $(\beta_i|\tau_i)$ and the jth column. For example,

$$\chi_1^{SM}(C_{2x}|\tfrac{1}{2}\tfrac{1}{2}) = 1, T = e^{ik_{SM}\cdot\tau_i} = e^{i\pi\alpha}$$

and

$$\chi_4^{SM}(C_{2x}|\tfrac{1}{2}\tfrac{1}{2}) = -1, TJ = -\sqrt{-1}\, e^{i\pi\alpha}$$

One can also define $D_j^k(\beta_i)$ from $D_j^k(\beta_i) = D_j^k(\beta_i|\tau_i)e^{-ik\cdot\tau_i}$. For k vectors inside the Brillouin zone, the matrices $D_j^k(\beta_i)$ constitute an irreducible representation of the point group of the wave vector k. The phase factors $e^{-ik\cdot\tau_i}$ for all k vectors of each nonsymmorphic layer group are given in Table 28. For example, from Table 28, group 43, we have:

Group 43	$p2/a2/b2_1/a$		$1 = (0,\frac{1}{2})$		$2 = (\frac{1}{2},\frac{1}{2})$		$3 = (\frac{1}{2},0)$	
	DT	SM	X	Y	S	D	C	V
,1	T^*	1	1	$-J$	$-J$	T^*	$-J$	
,2	T^*	T^*	$-J$	$-J$	-1	T^*	T^*	
,3	1	T^*	$-J$	1	$-J$	$-J$	T^*	T^*

Following the group number and symbol are the three nonprimitive translations appearing among the elements of this group. The k vectors are denoted in the second row, and their components are listed in Table 24. The left-hand column indexes the nonprimitive translations τ_i. The phase factor $e^{-ik\cdot\tau_i}$ is given at the intersection of the ith row and the kth column. If the phase factor is not a constant, it is denoted by the $T^* = e^{ik\cdot\tau_i}$. For example,

$$e^{-ik_{DT}\cdot\tau_1} = e^{-i(0,\alpha)\cdot(0,\frac{1}{2})2\pi} = e^{-i\pi\alpha}$$

TABLE 28

Group 5 $pb11$ $1 = (0,\tfrac{1}{2})$

	A	B	Y	F	FA
,1	J	1	-J	T*	T*

Group 7 $p2/b11$ $1 = (0,\tfrac{1}{2})$

	A	B	Y	F
,1	J	1	-J	T*

Group 9 $p112_1$ $1 = (0,\tfrac{1}{2})$

	DT	X	Y	S	D	TA	DA
,1	T*	1	-J	-J	T*	T*	T*

Group 12 $p11a$ $1 = (\tfrac{1}{2},0)$

	SM	X	Y	S	C	SN	CA
,1	T*	-J	1	-J	T*	T*	T*

Group 15 $p112_1/m$ $1 = (0,\tfrac{1}{2})$

	SM	X	Y	S	D	C	DT
,1	1	1	-J	-J	T*	-J	T*

Group 17 $p112/a$ $1 = (\tfrac{1}{2},0)$

	DT	SM	X	Y	S	D	C
,1	1	T*	-J	1	-J	-J	T*

Group 18 $p112_1/a$ $1 = (\tfrac{1}{2},\tfrac{1}{2})$

	DT	SM	X	Y	S	D	C
,1	T*	T*	-J	-J	-1	T*	T*

Group 20 $p222_1$ $1 = (0,\tfrac{1}{2})$

	DT	SM	X	Y	S	D	C
,1	T*	1	1	-J	-J	T*	-J

(continued)

TABLE 28
(continued)

Group 21   p22$_1$2$_1$ 1 = (½,½)

	DT	SM	X	Y	S	D	C
,1	T*	T*	-J	-J	-1	T*	T*

Group 25   pm2$_1$a   1 = (½,0)

	SM	X	Y	S	C	SN	CA
,1	T*	-J	1	-J	T*	T*	T*

Group 26   pbm2$_1$ 1 = (0,½)

	DT	SM	X	Y	S	D	C	V	VA	DA	TA
,1	T*	1	1	-J	-J	T*	-J	T*	T*	T*	T*

Group 27   pbb2   1 = (0,½)

	DT	SM	X	Y	S	D	C	V	VA	DA	TA
,1	T*	1	1	-J	-J	T*	-J	T*	T*	T*	T*

Group 28   p2ma   1 = (½,0)

	DT	SM	X	Y	S	D	C
,1	1	T*	-J	1	-J	-J	T*

Group 29   pam2   1 = (½,0)

	DT	SM	X	Y	S	D	C	V	VA	DA	TA
,1	1	T*	-J	1	-J	-J	T*	T*	T*	-J	1

Group 30   pab2$_1$ 1 = (0,½);   2 = (½,½);   3 = (½,0)

	DT	SM	X	Y	S	D	C	V	VA	DA	TA
,1	T*		1	-J	-J	T*				T*	T*
,2	T*			-J	-J	-1	T*			T*	T*
,3	1	T*	-J	1	-J	-J	T*	T*	T*	-J	1

(continued)

TABLE 28
(continued)

Group 31   pnb2   l = (½,½)

	DT	SM	X	Y	S	D	C	V	VA	DA	TA
,1	T*	T*	-J	-J	-1	T*	T*	T*	T*	T*	T*

Group 32   pnm2₁   l = (½,½)

	DT	SM	X	Y	S	D	C	V	VA	DA	TA
,1	T*	T*	-J	-J	-1	T*	T*	T*	T*	T*	T*

Group 33   p2ba   l = (½,½)

	DT	SM	X	Y	S	C	D
,1	T*	T*	-J	-J	-1	T*	T*

Group 36   cam2   l = (-½,½)

	DT	SM	YA	YB	S	C	F	P	DA	FA	PA
,1	T*	1	1	-1	-J	-1	T*	T*	T*	T*	T*

Group 38   p2/a2/m2/a   l = (½,0)

	DT	SM	X	Y	S	D	C	V
,1	1	T*	-J	1	-J	-J	T*	T*

Group 39   p2/n2/b2/a   l = (½,½)

	DT	SM	X	Y	S	D	C	V
,1	T*	T*	-J	-J	-1	T*	T*	T*

Group 40   p2/m2₁/m2/a   l = (½,0)

	DT	SM	X	Y	S	D	C
,1	1	T*	-J	1	-J	-J	T*

Group 41   p2/a2₁/m2/m   l = (½,0)

	DT	SM	X	Y	S	D	C	V
,1	1	T*	-J	1	-J	-J	T*	T*

(continued)

TABLE 28
(continued)

Group 42 $p2/n2/m2_1/a$ $1 = (\frac{1}{2},\frac{1}{2})$

	DT	SM	X	Y	S	D	C	V
,1	T*	T*	-J	-J	-1	T*	T*	T*

Group 43 $p2/a2/b2_1/a$ $1 = (0,\frac{1}{2})$; $2 = (\frac{1}{2},\frac{1}{2})$; $3 = (\frac{1}{2},0)$.

	DT	SM	X	Y	S	D	C	V
,1	T*	1	1	-J	-J	T*	-J	
,2	T*	T*	-J	-J	-1	T*	T*	
,3	1	T*	-J	1	-J	-J	T*	T*

Group 44 $p2/m2_1/b2_1/a$ $1 = (\frac{1}{2},\frac{1}{2})$

	DT	SM	X	Y	S	D	C
,1	T*	T*	-J	-J	-1	T*	T*

Group 45 $p2/a2_1/b2_1/m$ $1 = (0,\frac{1}{2})$; $2 = (\frac{1}{2},0)$; $3 = (\frac{1}{2},\frac{1}{2})$

	DT	SM	X	Y	S	D	C	V
,1	T*	1	1	-J	-J	T*	-J	
,2	1	T*	-J	1	-J	-J	T*	T*
,3	T*	T*	-J	-J	-1	T*	T*	

Group 46 $p2/n2_1/m2_1/m$ $1 = (\frac{1}{2},\frac{1}{2})$

	DT	SM	X	Y	S	D	C	V
,1	T*	T*	-J	-J	-1	T*	T*	T*

Group 48 $c2/a2/m2/m$ $1 = (\frac{1}{2},\frac{1}{2})$

	DT	SM	YA	YB	S	C	F	P
,1	1	T*	-1	1	-J	T*	-1	T*

Group 52 $p4/n$ $1 = (\frac{1}{2},\frac{1}{2})$

	SM	DT	M	X	Y	D	DA
,1	T*	T*	-1	-J	T*	T*	T*

(continued)

TABLE 28
(continued)

Group 54   p42$_1$2   1 = ($\frac{1}{2}$,$\frac{1}{2}$)

	DT	M	X	Y
,1	T*	-1	-J	T*

Group 56   p4bm   1 = ($\frac{1}{2}$,$\frac{1}{2}$)

	SM	DT	M	X	Y.
,1	T*	T*	-1	-J	T*

Group 58   p$\overline{4}$2$_1$m   1 = ($\frac{1}{2}$,$\frac{1}{2}$)

	SM	DT	M	X	Y
,1	T*	T*	-1	-J	T*

Group 60   p$\overline{4}$b2   1 = ($\frac{1}{2}$,$\frac{1}{2}$)

	SM	DT	M	X	Y
,1	T*	T*	-1	-J	T*

Group 62   p4/n2/b2/m   1 = ($\frac{1}{2}$,$\frac{1}{2}$)

	SM	DT	M	X	Y	D
,1	T*	T*	-1	-J	T*	T*

Group 63   p4/m2$_1$/b2/m   1 = ($\frac{1}{2}$,$\frac{1}{2}$)

	SM	DT	M	X	Y
,1	T*	T*	-1	-J	T*

Group 64   p4/n2$_1$/m2/m   1 = ($\frac{1}{2}$,$\frac{1}{2}$)

	SM	DT	M	X	Y	D
,1	T*	T*	-1	-J	T*	T*

7. Compatibility Relations

Relationships between the irreducible representations at different k vectors are called compatibility relations.[38] Compatibility relations provide information to correctly label, e.g., branches of phonon dispersion curves which have split from a single curve.[39] In Table 29 (see Appendix II) we have tabulated all compatibility relations for the eighty layer groups. For example, for group 40, $p2/m2_1/m2/a$, we have the following subtable of compatibility relations:

X	1	2	3	4
D	1 4	2 3	5	5
SM	1 3	2 4	5	5
V	(2)1	(2)2	3 4	3 4

This subtable gives the relationships between the irreducible representations D_i^X, i = 1,2,3,4, and the irreducible representations of neighboring k vectors in the corresponding representation domain of this layer group. The neighboring k vectors are listed in the left-hand column and the irreducible representation index i of D_i^X in the top row. The indices j of the irreducible representations D_j^k into which D_i^X splits are given at the intersections of the kth row and ith column. The multiplicity of a given irreducible representation D_j^k, if greater than one, is given in parentheses. We have from this subtable, for example, that $D_1^X \rightarrow D_1^{SM} + D_3^{SM}$ and $D_2^X \rightarrow 2D_2^V$.

8. The Seventeen Plane Groups

Seventeen of the eighty layer groups are mathematically isomorphic to the seventeen plane groups. Consequently, the irreducible representations and compatibility relations for the seventeen plane groups are identical with the irreducible representations and compatibility relations of these seventeen layer groups. In Table 30 we list the seventeen plane groups in the standard numbering and notation

TABLE 30

Plane Group	Layer Group
1	1
2	3
3	11
4	12
5	13
6	23
7	28
8	33
9	34
10	49
11	55
12	56
13	65
14	69
15	70
16	73
17	77

of the International Tables for Crystallography.[1] Alongside each plane group we list the corresponding layer group.

References

1. International Tables for Crystallography, Vol. A: Space-Group Symmetry. Edited by T. Hahn (D. Reidel, Dordrecht, 1983). See also International Tables for X-Ray Crystallography, Vol. 1, edited by N.F.M. Henry and K. Lonsdale (Kynoch Press, Birmingham, 1969).

2. A.P. Cracknell, Thin Solid Films, 21, 107 (1974).

3. D.B. Litvin, Thin Solid Films, 106, 203 (1983).

4. D.K. Faddeyev, Tables of the Principal Unitary Representations of Fedorov Groups (Pergamon, Oxford, 1964).

5. O.V. Kovalev, Irreducible Representations of the Space Groups (Gordon and Breach, New York, 1965).

6. S.C. Miller and W.F. Love, Tables of Irreducible Representations of Space Groups and Corepresentations of Magnetic Groups (Pruett, Boulder, 1967).

7. J. Zak, A. Casher, M. Gluk, and Y. Gur, The Irreducible Representaions of Space Groups (Benjamin, New York, 1969).

8. C.J. Bradley and A.P. Cracknell, The Mathematical Theory of Symmetry in Solids: Representation Theory for Point Groups and Space Groups (Oxford University Press, 1972).

9. A.P. Cracknell, B.L. Davies, S.C. Miller, and W.F. Love, Kronecker Product Tables, Vol. 1 (IFI/Plenum, New York, 1979).

10. A. Speiser, <u>Die Theorie der Gruppen von Endlicher Ordnung</u>, 2nd Edition (Springer, Berlin, 1927).

11. C. Hermann, Z. Krist., <u>69</u>, 250 (1928).

12. C. Hermann, Z. Krist., <u>69</u>, 533 (1928).

13. E. Alexander and K. Herrmann, Z. Krist., <u>69</u>, 285 (1928).

14. E. Alexander and K. Herrmann, Z. Krist., <u>70</u>, 328 (1929).

15. L. Weber, Z. Krist., <u>70</u>, 309 (1929).

16. E.A. Wood, J. Appl. Phys., <u>35</u>, 1306 (1964).

17. H. Brown, R. Bulow, J. Neubuser, H. Wondratschek, and H. Zessenhaus, <u>Crystallographic Groups in Four-Dimensional Space</u> (John Wiley, New York, 1978).

18. W.T. Holser, Z. Krist., <u>110</u>, 249 (1958).

19. W.T. Holser, Z. Krist., <u>110</u>, 255 (1958).

20. V. Janovec, Ferroelectrics, <u>35</u>, 105 (1981).

21. L.H. Germer, Ann. N.Y. Acad. Sci., <u>101</u>, 599 (1963).

22. A.V. MacRae, Science, <u>139</u>, 379 (1963).

23. F. Jona, J.V. Strozier, and W.S. Wang, Rep. Prog. Phys., <u>45</u>, 527 (1982).

24. R. Zallen, M.L. Slade, and A.T. Ward, Phys. Rev., <u>B13</u>, 4257 (1971).

25. Z.V. Popovic and H.J. Stulz, Phys. Stat. Sol., <u>B105</u>, 337 (1981).

26. E.A. Wood, Bell Telephone Tech. J., <u>43</u>, 541 (1964).

27. E.A. Wood, Bell Telephone Technical Publications, Monograph 4680 (1964).

28. J. Bohm and K. Dornberger-Schiff, Acta Cryst., <u>23</u>, 913 (1967).

29. A.V. Shubnikov and V.A. Koptsik, <u>Symmetry in Science and Art</u> (Plenum, New York, 1974).

30. K. Dornberger-Schiff, Acta Cryst., <u>9</u>, 593 (1956).

31. W. Cochran, Acta Cryst., <u>5</u>, 630 (1952).

32. N.V. Belov and T.N. Tarkhova, Kristallografiya, <u>1</u>, 4 (1956).

33. N.V. Belov, Sov. Phys.-Cryst., <u>4</u>, 730 (1959).

34. G. Chapuis, Dissertation, University of Zurich, unpublished (1966).

35. K.J. Kohler, Dissertation, Aachen, unpublished (1977).

36. F. Seitz, Ann. Math., <u>37</u>, 17 (1936).

37. G.F. Koster, <u>Space Groups and Their Representations</u> (Academic Press, New York, 1957).

38. L.P. Bouckaert, R. Smoluchowski, and E.P. Wigner, Phys. Rev., <u>50</u>, 58 (1936).

39. D.B. Litvin, J. Phys., <u>C17</u>, L37 (1984).

APPENDIX I

Character Tables of the Irreducible Representations

of the Eighty Layer Groups

GROUP 2   P1-   GROUP 3   P211

          GM   GM
                    1+   1-   2+   2-   1   2   3   4

    1   1   1   1   1   1   1   1   1   1
   49   1   1   -1   -1   49   1   1   -1   -1
   25   1   -1   1   -1   4   1   -1   J   -J
   73   1   -1   -1   1   52   1   -1   -J   J

          A   A
                    1+   1-   2+   2-   1   2   3   4

    1   1   1   1   1   1   1   1   1   1
   49   1   1   -1   -1   49   1   1   -1   -1
   25   1   -1   1   -1   4   1   -1   J   -J
   73   1   -1   -1   1   52   1   -1   -J   J

          B   B
                    1+   1-   2+   2-   1   2   3   4

    1   1   1   1   1   1   1   1   1   1
   49   1   1   -1   -1   49   1   1   -1   -1
   25   1   -1   1   -1   4   1   -1   J   -J
   73   1   -1   -1   1   52   1   -1   -J   J

          Y   Y
                    1+   1-   2+   2-   1   2   3   4

    1   1   1   1   1   1   1   1   1   1
   49   1   1   -1   -1   49   1   1   -1   -1
   25   1   -1   1   -1   4   1   -1   J   -J
   73   1   -1   -1   1   52   1   -1   -J   J

GROUP 4   PM11

GM

	1	2	3	4
1	1	1	1	1
49	1	1	-1	-1
28	1	-1	J	-J
76	1	-1	-J	J

A

	1	2	3	4
1	1	1	1	1
49	1	1	-1	-1
28	1	-1	J	-J
76	1	-1	-J	J

B

	1	2	3	4
1	1	1	1	1
49	1	1	-1	-1
28	1	-1	J	-J
76	1	-1	-J	J

Y

	1	2	3	4
1	1	1	1	1
49	1	1	-1	-1
28	1	-1	J	-J
76	1	-1	-J	J

F

	1	2	3	4
1	1	1	1	1
49	1	1	-1	-1
28	1	-1	J	-J
76	1	-1	-J	J

FA

	1	2	3	4
1	1	1	1	1
49	1	1	-1	-1
28	1	-1	J	-J
76	1	-1	-J	J

GROUP 5   PB11

GM

	1	2	3	4
1	1	1	1	1
49	1	1	-1	-1
28,1	1	-1	J	-J
76,1	1	-1	-J	J

A

	1	2	3	4
1	1	1	1	1
49	1	1	-1	-1
28,1	J	-J	1	-1
76,1	J	-J	-1	1

B

	1	2	3	4
1	1	1	1	1
49	1	1	-1	-1
28,1	1	-1	J	-J
76,1	1	-1	-J	J

Y

	1	2	3	4
1	1	1	1	1
49	1	1	-1	-1
28,1	J	-J	1	-1
76,1	J	-J	-1	1

F

	1	2	3	4
1	1	1	1	1
49	1	1	-1	-1
28,1	1,T	-1,T	-1,TJ	1,TJ
76,1	1,T	-1,T	1,TJ	-1,TJ

FA

	1	2	3	4
1	1	1	1	1
49	1	1	-1	-1
28,1	1,T	-1,T	-1,TJ	1,TJ
76,1	1,T	-1,T	1,TJ	-1,TJ

GM

	1+	2+	1−	2−	3+	4+	3−	4−
1	1	1	1	1	1	1	1	1
49	1	1	1	1	−1	−1	−1	−1
4	1	−1	1	−1	J	−J	J	−J
28	1	−1	−1	1	J	−J	−J	J
73	1	1	−1	−1	−1	−1	1	1
52	1	−1	1	−1	−J	J	−J	J
76	1	−1	−1	1	−J	J	J	−J
25	1	1	−1	−1	1	1	−1	−1

A

	1+	2+	1−	2−	3+	4+	3−	4−
1	1	1	1	1	1	1	1	1
49	1	1	1	1	−1	−1	−1	−1
4	1	−1	1	−1	J	−J	J	−J
28	1	−1	−1	1	J	−J	−J	J
73	1	1	−1	−1	−1	−1	1	1
52	1	−1	1	−1	−J	J	−J	J
76	1	−1	−1	1	−J	J	J	−J
25	1	1	−1	−1	1	1	−1	−1

B

	1+	2+	1−	2−	3+	4+	3−	4−
1	1	1	1	1	1	1	1	1
49	1	1	1	1	−1	−1	−1	−1
4	1	−1	1	−1	J	−J	J	−J
28	1	−1	−1	1	J	−J	−J	J
73	1	1	−1	−1	−1	−1	1	1
52	1	−1	1	−1	−J	J	−J	J
76	1	−1	−1	1	−J	J	J	−J
25	1	1	−1	−1	1	1	−1	−1

Y

	1+	2+	1−	2−	3+	4+	3−	4−
1	1	1	1	1	1	1	1	1
49	1	1	1	1	−1	−1	−1	−1
4	1	−1	1	−1	J	−J	J	−J
28	1	−1	−1	1	J	−J	−J	J
73	1	1	−1	−1	−1	−1	1	1
52	1	−1	1	−1	−J	J	−J	J
76	1	−1	−1	1	−J	J	J	−J
25	1	1	−1	−1	1	1	−1	−1

F

	1	2	3	4
1	1	1	1	1
49	1	1	−1	−1
28	1	−1	J	−J
76	1	−1	−J	J

GM

	1+	2+	1-	2-	3+	4+	3-	4-
1	1	1	1	1	1	1	1	1
49	1	1	1	1	-1	-1	-1	-1
4,1	1	-1	1	-1	J	-J	J	-J
28,1	1	-1	-1	1	J	-J	-J	J
73	1	1	-1	-1	-1	-1	1	1
52,1	1	-1	1	-1	-J	J	-J	J
76,1	1	-1	-1	1	-J	J	J	-J
25	1	1	-1	-1	1	1	-1	-1

A

	1	2
1	2	2
49	2	-2
4,1 28,1 73		
52,1 76,1 25	0	0

B

	1+	2+	1-	2-	3+	4+	3-	4-
1	1	1	1	1	1	1	1	1
49	1	1	1	1	-1	-1	-1	-1
4,1	1	-1	1	-1	J	-J	J	-J
28,1	1	-1	-1	1	J	-J	-J	J
73	1	1	-1	-1	-1	-1	1	1
52,1	1	-1	1	-1	-J	J	-J	J
76,1	1	-1	-1	1	-J	J	J	-J
25	1	1	-1	-1	1	1	-1	-1

Y

	1	2
1	2	2
49	2	-2
4,1 28,1 73		
52,1 76,1 25	0	0

F

	1	2	3	4
1	1	1	1	1
49	1	1	-1	-1
28,1	1,T	-1,T	-1,TJ	1,TJ
76,1	1,T	-1,T	1,TJ	-1,TJ

GROUP 8 — P112

GM

	1	2	3	4
1	1	1	1	1
49	1	1	-1	-1
3	1	-1	J	-J
51	1	-1	-J	J

DT

	1	2	3	4
1	1	1	1	1
49	1	1	-1	-1
3	1	-1	J	-J
51	1	-1	-J	J

X

	1	2	3	4
1	1	1	1	1
49	1	1	-1	-1
3	1	-1	J	-J
51	1	-1	-J	J

Y

	1	2	3	4
1	1	1	1	1
49	1	1	-1	-1
3	1	-1	J	-J
51	1	-1	-J	J

S

	1	2	3	4
1	1	1	1	1
49	1	1	-1	-1
3	1	-1	J	-J
51	1	-1	-J	J

D

	1	2	3	4
1	1	1	1	1
49	1	1	-1	-1
3	1	-1	J	-J
51	1	-1	-J	J

TA

	1	2	3	4
1	1	1	1	1
49	1	1	-1	-1
3	1	-1	J	-J
51	1	-1	-J	J

DA

	1	2	3	4
1	1	1	1	1
49	1	1	-1	-1
3	1	-1	J	-J
51	1	-1	-J	J

GROUP 9 — P112(1)

GM

	1	2	3	4
1	1	1	1	1
49	1	1	-1	-1
3,1	1	-1	J	-J
51,1	1	-1	-J	J

DT

	1	2	3	4
1	1	1	1	1
49	1	1	-1	-1
3,1	1,T	-1,T	-1,TJ	1,TJ
51,1	1,T	-1,T	1,TJ	-1,TJ

X

	1	2	3	4
1	1	1	1	1
49	1	1	-1	-1
3,1	1	-1	J	-J
51,1	1	-1	-J	J

Y

	1	2	3	4
1	1	1	1	1
49	1	1	-1	-1
3,1	J	-J	1	-1
51,1	J	-J	-1	1

S

	1	2	3	4
1	1	1	1	1
49	1	1	-1	-1
3,1	J	-J	1	-1
51,1	J	-J	-1	1

D

	1	2	3	4
1	1	1	1	1
49	1	1	-1	-1
3,1	1,T	-1,T	-1,TJ	1,TJ
51,1	1,T	-1,T	1,TJ	-1,TJ

TA

	1	2	3	4
1	1	1	1	1
49	1	1	-1	-1
3,1	1,T	-1,T	-1,TJ	1,TJ
51,1	1,T	-1,T	1,TJ	-1,TJ

DA

	1	2	3	4
1	1	1	1	1
49	1	1	-1	-1
3,1	1,T	-1,T	-1,TJ	1,TJ
51,1	1,T	-1,T	1,TJ	-1,TJ

GROUP 10 — C112

GM

	1	2	3	4
1	1	1	1	1
49	1	1	-1	-1
3	1	-1	J	-J
51	1	-1	-J	J

DT

	1	2	3	4
1	1	1	1	1
49	1	1	-1	-1
3	1	-1	J	-J
51	1	-1	-J	J

VA

	1	2	3	4
1	1	1	1	1
49	1	1	-1	-1
3	1	-1	J	-J
51	1	-1	-J	J

YB

	1	2	3	4
1	1	1	1	1
49	1	1	-1	-1
3	1	-1	J	-J
51	1	-1	-J	J

S

	1	2	3	4
1	1	1	1	1
49	1	1	-1	-1
3	1	-1	J	-J
51	1	-1	-J	J

F

	1	2	3	4
1	1	1	1	1
49	1	1	-1	-1
3	1	-1	J	-J
51	1	-1	-J	J

DA

	1	2	3	4
1	1	1	1	1
49	1	1	-1	-1
3	1	-1	J	-J
51	1	-1	-J	J

FA

	1	2	3	4
1	1	1	1	1
49	1	1	-1	-1
3	1	-1	J	-J
51	1	-1	-J	J

GROUP 11 — P11M

GM

	1	2	3	4
1	1	1	1	1
49	1	1	-1	-1
27	1	-1	J	-J
75	1	-1	-J	J

SM

	1	2	3	4
1	1	1	1	1
49	1	1	-1	-1
27	1	-1	J	-J
75	1	-1	-J	J

X

	1	2	3	4
1	1	1	1	1
49	1	1	-1	-1
27	1	-1	J	-J
75	1	-1	-J	J

Y

	1	2	3	4
1	1	1	1	1
49	1	1	-1	-1
27	1	-1	J	-J
75	1	-1	-J	J

S

	1	2	3	4
1	1	1	1	1
49	1	1	-1	-1
27	1	-1	J	-J
75	1	-1	-J	J

C

	1	2	3	4
1	1	1	1	1
49	1	1	-1	-1
27	1	-1	J	-J
75	1	-1	-J	J

SN

	1	2	3	4
1	1	1	1	1
49	1	1	-1	-1
27	1	-1	J	-J
75	1	-1	-J	J

CA

	1	2	3	4
1	1	1	1	1
49	1	1	-1	-1
27	1	-1	J	-J
75	1	-1	-J	J

GROUP 12 — P11A

GM

	1	2	3	4
1	1	1	1	1
49	1	1	-1	-1
27,1	1	-1	J	-J
75,1	1	-1	-J	J

SM

	1	2	3	4
1	1	1	1	1
49	1	1	-1	-1
27,1	1,T	-1,T	-1,TJ	1,TJ
75,1	1,T	-1,T	1,TJ	-1,TJ

X

	1	2	3	4
1	1	1	1	1
49	1	1	-1	-1
27,1	J	-J	1	-1
75,1	J	-J	-1	1

Y

	1	2	3	4
1	1	1	1	1
49	1	1	-1	-1
27,1	1	-1	J	-J
75,1	1	-1	-J	J

S

	1	2	3	4
1	1	1	1	1
49	1	1	-1	-1
27,1	J	-J	1	-1
75,1	J	-J	-1	1

C

	1	2	3	4
1	1	1	1	1
49	1	1	-1	-1
27,1	1,T	-1,T	-1,TJ	1,TJ
75,1	1,T	-1,T	1,TJ	-1,TJ

SN

	1	2	3	4
1	1	1	1	1
49	1	1	-1	-1
27,1	1,T	-1,T	-1,TJ	1,TJ
75,1	1,T	-1,T	1,TJ	-1,TJ

CA

	1	2	3	4
1	1	1	1	1
49	1	1	-1	-1
27,1	1,T	-1,T	-1,TJ	1,TJ
75,1	1,T	-1,T	1,TJ	-1,TJ

GROUP 13 — C11M

GM

	1	2	3	4
1	1	1	1	1
49	1	1	-1	-1
27	1	-1	J	-J
75	1	-1	-J	J

SM

	1	2	3	4
1	1	1	1	1
49	1	1	-1	-1
27	1	-1	J	-J
75	1	-1	-J	J

YA

	1	2	3	4
1	1	1	1	1
49	1	1	-1	-1
27	1	-1	J	-J
75	1	-1	-J	J

YB

	1	2	3	4
1	1	1	1	1
49	1	1	-1	-1
27	1	-1	J	-J
75	1	-1	-J	J

C

	1	2	3	4
1	1	1	1	1
49	1	1	-1	-1
27	1	-1	J	-J
75	1	-1	-J	J

CA

	1	2	3	4
1	1	1	1	1
49	1	1	-1	-1
27	1	-1	J	-J
75	1	-1	-J	J

SN

	1	2	3	4
1	1	1	1	1
49	1	1	-1	-1
27	1	-1	J	-J
75	1	-1	-J	J

57

GM

	1+	2+	1-	2-	3+	4+	3-	4-
1	1	1	1	1	1	1	1	1
49	1	1	1	1	-1	-1	-1	-1
3	1	-1	1	-1	J	-J	J	-J
27	1	-1	-1	1	J	-J	-J	J
73	1	1	-1	-1	-1	-1	1	1
51	1	-1	1	-1	-J	J	-J	J
75	1	-1	-1	1	-J	J	J	-J
25	1	1	-1	-1	1	1	-1	-1

DT

	1	2	3	4
1	1	1	1	1
49	1	1	-1	-1
3	1	-1	J	-J
51	1	-1	-J	J

SM

	1	2	3	4
1	1	1	1	1
49	1	1	-1	-1
27	1	-1	J	-J
75	1	-1	-J	J

X

	1+	2+	1-	2-	3+	4+	3-	4-
1	1	1	1	1	1	1	1	1
49	1	1	1	1	-1	-1	-1	-1
3	1	-1	1	-1	J	-J	J	-J
27	1	-1	-1	1	J	-J	-J	J
73	1	1	-1	-1	-1	-1	1	1
51	1	-1	1	-1	-J	J	-J	J
75	1	-1	-1	1	-J	J	J	-J
25	1	1	-1	-1	1	1	-1	-1

Y

	1+	2+	1-	2-	3+	4+	3-	4-
1	1	1	1	1	1	1	1	1
49	1	1	1	1	-1	-1	-1	-1
3	1	-1	1	-1	J	-J	J	-J
27	1	-1	-1	1	J	-J	-J	J
73	1	1	-1	-1	-1	-1	1	1
51	1	-1	1	-1	-J	J	-J	J
75	1	-1	-1	1	-J	J	J	-J
25	1	1	-1	-1	1	1	-1	-1

S

	1+	2+	1-	2-	3+	4+	3-	4-
1	1	1	1	1	1	1	1	1
49	1	1	1	1	-1	-1	-1	-1
3	1	-1	1	-1	J	-J	J	-J
27	1	-1	-1	1	J	-J	-J	J
73	1	1	-1	-1	-1	-1	1	1
51	1	-1	1	-1	-J	J	-J	J
75	1	-1	-1	1	-J	J	J	-J
25	1	1	-1	-1	1	1	-1	-1

D

	1	2	3	4
1	1	1	1	1
49	1	1	-1	-1
3	1	-1	J	-J
51	1	-1	-J	J

C

	1	2	3	4
1	1	1	1	1
49	1	1	-1	-1
27	1	-1	J	-J
75	1	-1	-J	J

GM

	1+	2+	1-	2-	3+	4+	3-	4-
1	1	1	1	1	1	1	1	1
49	1	1	1	1	-1	-1	-1	-1
3,1	1	-1	1	-1	J	-J	J	-J
27,1	1	-1	-1	1	J	-J	-J	J
73	1	1	-1	-1	-1	-1	1	1
51,1	1	-1	1	-1	-J	J	-J	J
75,1	1	-1	-1	1	-J	J	J	-J
25	1	1	-1	-1	1	1	-1	-1

SM

	1	2	3	4
1	1	1	1	1
49	1	1	-1	-1
27,1	1	-1	J	-J
75,1	1	-1	-J	J

X

	1+	2+	1-	2-	3+	4+	3-	4-
1	1	1	1	1	1	1	1	1
49	1	1	1	1	-1	-1	-1	-1
3,1	1	-1	1	-1	J	-J	J	-J
27,1	1	-1	-1	1	J	-J	-J	J
73	1	1	-1	-1	-1	-1	1	1
51,1	1	-1	1	-1	-J	J	-J	J
75,1	1	-1	-1	1	-J	J	J	-J
25	1	1	-1	-1	1	1	-1	-1

Y

	1	2
1	2	2
49	2	-2
3,1 27,1 73		
51,1 75,1 25	0	0

S

	1	2
1	2	2
49	2	-2
3,1 27,1 73		
51,1 75,1 25	0	0

D

	1	2	3	4
1	1	1	1	1
49	1	1	-1	-1
3,1	1,T	-1,T	-1,TJ	1,TJ
51,1	1,T	-1,T	1,TJ	-1,TJ

C

	1	2	3	4
1	1	1	1	1
49	1	1	-1	-1
27,1	1	-1	J	-J
75,1	1	-1	-J	J

DT

	1	2	3	4
1	1	1	1	1
49	1	1	-1	-1
3,1	1,T	-1,T	-1,TJ	1,TJ
51,1	1,T	-1,T	1,TJ	-1,TJ

GM

	1+	2+	1-	2-	3+	4+	3-	4-
1	1	1	1	1	1	1	1	1
49	1	1	1	1	-1	-1	-1	-1
3	1	-1	1	-1	J	-J	J	-J
27	1	-1	-1	1	J	-J	-J	J
73	1	1	-1	-1	-1	-1	1	1
51	1	-1	1	-1	-J	J	-J	J
75	1	-1	-1	1	-J	J	J	-J
25	1	1	-1	-1	1	1	-1	-1

DT

	1	2	3	4
1	1	1	1	1
49	1	1	-1	-1
3	1	-1	J	-J
51	1	-1	-J	J

SM

	1	2	3	4
1	1	1	1	1
49	1	1	-1	-1
27	1	-1	J	-J
75	1	-1	-J	J

VA

	1+	2+	1-	2-	3+	4+	3-	4-
1	1	1	1	1	1	1	1	1
49	1	1	1	1	-1	-1	-1	-1
3	1	-1	1	-1	J	-J	J	-J
27	1	-1	-1	1	J	-J	-J	J
73	1	1	-1	-1	-1	-1	1	1
51	1	-1	1	-1	-J	J	-J	J
75	1	-1	-1	1	-J	J	J	-J
25	1	1	-1	-1	1	1	-1	-1

YB

	1+	2+	1-	2-	3+	4+	3-	4-
1	1	1	1	1	1	1	1	1
49	1	1	1	1	-1	-1	-1	-1
3	1	-1	1	-1	J	-J	J	-J
27	1	-1	-1	1	J	-J	-J	J
73	1	1	-1	-1	-1	-1	1	1
51	1	-1	1	-1	-J	J	-J	J
75	1	-1	-1	1	-J	J	J	-J
25	1	1	-1	-1	1	1	-1	-1

S

	1	2	3	4
1	1	1	1	1
49	1	1	-1	-1
25	1	-1	1	-1
73	1	-1	-1	1

C

	1	2	3	4
1	1	1	1	1
49	1	1	-1	-1
27	1	-1	J	-J
75	1	-1	-J	J

F

	1	2	3	4
1	1	1	1	1
49	1	1	-1	-1
3	1	-1	J	-J
51	1	-1	-J	J

GM

	1+	2+	1-	2-	3+	4+	3-	4-
1	1	1	1	1	1	1	1	1
49	1	1	1	1	-1	-1	-1	-1
3,1	1	-1	1	-1	J	-J	J	-J
27,1	1	-1	-1	1	J	-J	-J	J
73	1	1	-1	-1	-1	-1	1	1
51,1	1	-1	1	-1	-J	J	-J	J
75,1	1	-1	-1	1	-J	J	J	-J
25	1	1	-1	-1	1	1	-1	-1

DT

	1	2	3	4
1	1	1	1	1
49	1	1	-1	-1
3,1	1	-1	J	-J
51,1	1	-1	-J	J

SM

	1	2	3	4
1	1	1	1	1
49	1	1	-1	-1
27,1	1,T	-1,T	-1,TJ	1,TJ
75,1	1,T	-1,T	1,TJ	-1,TJ

X

	1	2
1	2	2
49	2	-2
3,1 27,1 73		
51,1 75,1 25	0	0

Y

	1+	2+	1-	2-	3+	4+	3-	4-
1	1	1	1	1	1	1	1	1
49	1	1	1	1	-1	-1	-1	-1
3,1	1	-1	1	-1	J	-J	J	-J
27,1	1	-1	-1	1	J	-J	-J	J
73	1	1	-1	-1	-1	-1	1	1
51,1	1	-1	1	-1	-J	J	-J	J
75,1	1	-1	-1	1	-J	J	J	-J
25	1	1	-1	-1	1	1	-1	-1

S

	1	2
1	2	2
49	2	-2
3,1 27,1 73		
51,1 75,1 25	0	0

D

	1	2	3	4
1	1	1	1	1
49	1	1	-1	-1
3,1	1	-1	J	-J
51,1	1	-1	-J	J

C

	1	2	3	4
1	1	1	1	1
49	1	1	-1	-1
27,1	1,T	-1,T	-1,TJ	1,TJ
75,1	1,T	-1,T	1,TJ	-1,TJ

GM

	1+	2+	1-	2-	3+	4+	3-	4-
1	1	1	1	1	1	1	1	1
49	1	1	1	1	-1	-1	-1	-1
3,1	1	-1	1	-1	J	-J	J	-J
27,1	1	-1	-1	1	J	-J	-J	J
73	1	1	-1	-1	-1	-1	1	1
51,1	1	-1	1	-1	-J	J	-J	J
75,1	1	-1	-1	1	-J	J	J	-J
25	1	1	-1	-1	1	1	-1	-1

DT

	1	2	3	4
1	1	1	1	1
49	1	1	-1	-1
3,1	1,T	-1,T	-1,TJ	1,TJ
51,1	1,T	-1,T	1,TJ	-1,TJ

SM

	1	2	3	4
1	1	1	1	1
49	1	1	-1	-1
27,1	1,T	-1,T	-1,TJ	1,TJ
75,1	1,T	-1,T	1,TJ	-1,TJ

X

	1	2
1	2	2
49	2	-2
3,1 27,1 73		
51,1 75,1 25	0	0

Y

	1	2
1	2	2
49	2	-2
3,1 27,1 73		
51,1 75,1 25	0	0

S

	1+	2+	1-	2-	3+	4+	3-	4-
1	1	1	1	1	1	1	1	1
49	1	1	1	1	-1	-1	-1	-1
3,1	J	-J	J	-J	1	-1	1	-1
27,1	J	-J	-J	J	1	-1	-1	1
73	1	1	-1	-1	-1	-1	1	1
51,1	J	-J	J	-J	-1	1	-1	1
75,1	J	-J	-J	J	-1	1	1	-1
25	1	1	-1	-1	1	1	-1	-1

D

	1	2	3	4
1	1	1	1	1
49	1	1	-1	-1
3,1	-1,TJ	1,TJ	-1,T	1,T
51,1	-1,TJ	1,TJ	1,T	-1,T

C

	1	2	3	4
1	1	1	1	1
49	1	1	-1	-1
27,1	1,TJ	-1,TJ	1,T	-1,T
75,1	1,TJ	-1,TJ	-1,T	1,T

GROUP 19   P222

GM

	1	2	3	4	5
1	1	1	1	1	2
49	1	1	1	1	-2
4 52	1	1	-1	-1	0
2 50	1	-1	1	-1	0
3 51	1	-1	-1	1	0

DT

	1	2	3	4
1	1	1	1	1
49	1	1	-1	-1
51	1	-1	J	-J
3	1	-1	-J	J

SM

	1	2	3	4
1	1	1	1	1
49	1	1	-1	-1
2	1	-1	J	-J
50	1	-1	-J	J

X

	1	2	3	4	5
1	1	1	1	1	2
49	1	1	1	1	-2
4 52	1	1	-1	-1	0
2 50	1	-1	1	-1	0
3 51	1	-1	-1	1	0

Y

	1	2	3	4	5
1	1	1	1	1	2
49	1	1	1	1	-2
4 52	1	1	-1	-1	0
2 50	1	-1	1	-1	0
3 51	1	-1	-1	1	0

S

	1	2	3	4	5
1	1	1	1	1	2
49	1	1	1	1	-2
4 52	1	1	-1	-1	0
2 50	1	-1	1	-1	0
3 51	1	-1	-1	1	0

C

	1	2	3	4
1	1	1	1	1
49	1	1	-1	-1
2	1	-1	J	-J
50	1	-1	-J	J

D

	1	2	3	4
1	1	1	1	1
49	1	1	-1	-1
51	1	-1	J	-J
3	1	-1	-J	J

GM

	1	2	3	4	5
1	1	1	1	1	2
49	1	1	1	1	-2
3,1 51,1	1	1	-1	-1	0
4 52	1	-1	1	-1	0
2,1 50,1	1	-1	-1	1	0

DT

	1	2	3	4
1	1	1	1	1
49	1	1	-1	-1
3,1	1,T	-1,T	-1,TJ	1,TJ
51,1	1,T	-1,T	1,TJ	-1,TJ

SM

	1	2	3	4
1	1	1	1	1
49	1	1	-1	-1
50,1	1	-1	J	-J
2,1	1	-1	-J	J

X

	1	2	3	4	5
1	1	1	1	1	2
49	1	1	1	1	-2
3,1 51,1	1	1	-1	-1	0
4 52	1	-1	1	-1	0
2,1 50,1	1	-1	-1	1	0

Y

	1	2	3	4	5
1	2	1	1	1	1
49	2	-1	-1	-1	-1
3,1	0	1	1	-1	-1
4	0	J	-J	J	-J
50,1	0	-J	J	J	-J
51,1	0	-1	-1	1	1
52	0	-J	J	-J	J
2,1	0	J	-J	-J	J

S

	1	2	3	4	5
1	2	1	1	1	1
49	2	-1	-1	-1	-1
3,1	0	1	1	-1	-1
4	0	J	-J	J	-J
50,1	0	-J	J	J	-J
51,1	0	-1	-1	1	1
52	0	-J	J	-J	J
2,1	0	J	-J	-J	J

D

	1	2	3	4
1	1	1	1	1
49	1	1	-1	-1
3,1	1,T	-1,T	-1,TJ	1,TJ
51,1	1,T	-1,T	1,TJ	-1,TJ

C

	1	2	3	4
1	1	1	1	1
49	1	1	-1	-1
50,1	1	-1	J	-J
2,1	1	-1	-J	J

GM

	1	2	3	4	5
1	1	1	1	1	2
49	1	1	1	1	-2
4 52	1	1	-1	-1	0
2,1 50,1	1	-1	1	-1	0
3,1 51,1	1	-1	-1	1	0

DT

	1	2	3	4
1	1	1	1	1
49	1	1	-1	-1
51,1	1,T	-1,T	-1,TJ	1,TJ
3,1	1,T	-1,T	1,TJ	-1,TJ

SM

	1	2	3	4
1	1	1	1	1
49	1	1	-1	-1
2,1	1,T	-1,T	-1,TJ	1,TJ
50,1	1,T	-1,T	1,TJ	-1,TJ

X

	1	2	3	4	5
1	2	1	1	1	1
49	2	-1	-1	-1	-1
4	0	J	J	-J	-J
2,1	0	1	-1	1	-1
51,1	0	J	-J	-J	J
52	0	-J	-J	J	J
50,1	0	-1	1	-1	1
3,1	0	-J	J	J	-J

Y

	1	2	3	4	5
1	2	1	1	1	1
49	2	-1	-1	-1	-1
4	0	J	J	-J	-J
2,1	0	J	-J	J	-J
51,1	0	-1	1	1	-1
52	0	-J	-J	J	J
50,1	0	-J	J	-J	J
3,1	0	1	-1	-1	1

S

	1	2	3	4	5
1	1	1	1	1	2
49	1	1	1	1	-2
4 52	1	1	-1	-1	0
2,1 50,1	J	-J	J	-J	0
3,1 51,1	J	-J	-J	J	0

D

	1	2	3	4
1	1	1	1	1
49	1	1	-1	-1
51,1	-1,TJ	1,TJ	-1,T	1,T
3,1	-1,TJ	1,TJ	1,T	-1,T

C

	1	2	3	4
1	1	1	1	1
49	1	1	-1	-1
2,1	-1,TJ	1,TJ	-1,T	1,T
50,1	-1,TJ	1,TJ	1,T	-1,T

GROUP 22 — C222

GM

	1	2	3	4	5
1	1	1	1	1	2
49	1	1	1	1	-2
4 52	1	1	-1	-1	0
2 50	1	-1	1	-1	0
3 51	1	-1	-1	1	0

DT

	1	2	3	4
1	1	1	1	1
49	1	1	-1	-1
51	1	-1	J	-J
3	1	-1	-J	J

SM

	1	2	3	4
1	1	1	1	1
49	1	1	-1	-1
2	1	-1	J	-J
50	1	-1	-J	J

YA

	1	2	3	4	5
1	1	1	1	1	2
49	1	1	1	1	-2
4 52	1	1	-1	-1	0
2 50	1	-1	1	-1	0
3 51	1	-1	-1	1	0

YB

	1	2	3	4	5
1	1	1	1	1	2
49	1	1	1	1	-2
4 52	1	1	-1	-1	0
2 50	1	-1	1	-1	0
3 51	1	-1	-1	1	0

S

	1	2	3	4
1	1	1	1	1
49	1	1	-1	-1
4	1	-1	J	-J
52	1	-1	-J	J

C

	1	2	3	4
1	1	1	1	1
49	1	1	-1	-1
2	1	-1	J	-J
50	1	-1	-J	J

F

	1	2	3	4
1	1	1	1	1
49	1	1	-1	-1
51	1	-1	J	-J
3	1	-1	-J	J

GROUP 23 — P2MM

GM

	1	2	3	4	5
1	1	1	1	1	2
49	1	1	1	1	-2
4 52	1	1	-1	-1	0
26 74	1	-1	1	-1	0
27 75	1	-1	-1	1	0

DT

	1	2	3	4
1	1	1	1	1
49	1	1	-1	-1
26	1	-1	J	-J
74	1	-1	-J	J

SM

	1	2	3	4
1	1	1	1	1
49	1	1	-1	-1
75	1	-1	J	-J
27	1	-1	-J	J

X

	1	2	3	4	5
1	1	1	1	1	2
49	1	1	1	1	-2
4 52	1	1	-1	-1	0
26 74	1	-1	1	-1	0
27 75	1	-1	-1	1	0

V

	1	2	3	4	5
1	1	1	1	1	2
49	1	1	1	1	-2
4 52	1	1	-1	-1	0
26 74	1	-1	1	-1	0
27 75	1	-1	-1	1	0

S

	1	2	3	4	5
1	1	1	1	1	2
49	1	1	1	1	-2
4 52	1	1	-1	-1	0
26 74	1	-1	1	-1	0
27 75	1	-1	-1	1	0

C

	1	2	3	4
1	1	1	1	1
49	1	1	-1	-1
75	1	-1	J	-J
27	1	-1	-J	J

D

	1	2	3	4
1	1	1	1	1
49	1	1	-1	-1
26	1	-1	J	-J
74	1	-1	-J	J

PMM2

GM

	1	2	3	4	5
1	1	1	1	1	2
49	1	1	1	1	-2
3 51	1	1	-1	-1	0
26 74	1	-1	1	-1	0
28 76	1	-1	-1	1	0

DT

	1	2	3	4	5
1	1	1	1	1	2
49	1	1	1	1	-2
3 51	1	1	-1	-1	0
26 74	1	-1	1	-1	0
28 76	1	-1	-1	1	0

SM

	1	2	3	4
1	1	1	1	1
49	1	1	-1	-1
76	1	-1	J	-J
28	1	-1	-J	J

X

	1	2	3	4	5
1	1	1	1	1	2
49	1	1	1	1	-2
3 51	1	1	-1	-1	0
26 74	1	-1	1	-1	0
28 76	1	-1	-1	1	0

Y

	1	2	3	4	5
1	1	1	1	1	2
49	1	1	1	1	-2
3 51	1	1	-1	-1	0
26 74	1	-1	1	-1	0
28 76	1	-1	-1	1	0

S

	1	2	3	4	5
1	1	1	1	1	2
49	1	1	1	1	-2
3 51	1	1	-1	-1	0
26 74	1	-1	1	-1	0
28 76	1	-1	-1	1	0

D

	1	2	3	4	5
1	1	1	1	1	2
49	1	1	1	1	-2
3 51	1	1	-1	-1	0
26 74	1	-1	1	-1	0
28 76	1	-1	-1	1	0

C

	1	2	3	4
1	1	1	1	1
49	1	1	-1	-1
76	1	-1	J	-J
28	1	-1	-J	J

V

	1	2	3	4
1	1	1	1	1
49	1	1	-1	-1
28	1	-1	J	-J
76	1	-1	-J	J

VA

	1	2	3	4
1	1	1	1	1
49	1	1	-1	-1
28	1	-1	-J	J
76	1	-1	J	-J

TA

	1	2	3	4	5
1	1	1	1	1	2
49	1	1	1	1	-2
3 51	1	1	-1	-1	0
26 74	1	-1	1	-1	0
28 76	1	-1	-1	1	0

DA

	1	2	3	4	5
1	1	1	1	1	2
49	1	1	1	1	-2
3 51	1	1	-1	-1	0
26 74	1	-1	1	-1	0
28 76	1	-1	-1	1	0

GM

	1	2	3	4	5
1	1	1	1	1	2
49	1	1	1	1	-2
2,1 50,1	1	1	-1	-1	0
28 76	1	-1	1	-1	0
27,1 75,1	1	-1	-1	1	0

DT

	1	2	3	4
1	1	1	-1	1
49	1	1	-1	-1
28	1	-1	J	-J
76	1	-1	-J	J

SM

	1	2	3	4	5
1	1	1	1	1	2
49	1	1	1	1	-2
2,1 50,1	1,T	1,T	-1,T	-1,T	0
28 76	1	-1	1	-1	0
27,1 75,1	1,T	-1,T	-1,T	1,T	0

X

	1	2	3	4	5
1	1	1	1	1	2
49	1	1	1	1	-2
2,1 50,1	J	J	-J	-J	0
28 76	1	-1	1	-1	0
27,1 75,1	J	-J	-J	J	0

Y

	1	2	3	4	5
1	1	1	1	1	2
49	1	1	1	1	-2
2,1 50,1	1	1	-1	-1	0
28 76	1	-1	1	-1	0
27,1 75,1	1	-1	-1	1	0

S

	1	2	3	4	5
1	1	1	1	1	2
49	1	1	1	1	-2
2,1 50,1	J	J	-J	-J	0
28 76	1	-1	1	-1	0
27,1 75,1	J	-J	-J	J	0

D

	1	2	3	4
1	1	1	1	1
49	1	1	-1	-1
28	1	-1	J	-J
76	1	-1	-J	J

C

	1	2	3	4	5
1	1	1	1	1	2
49	1	1	1	1	-2
2,1 50,1	1,T	1,T	-1,T	-1,T	0
28 76	1	-1	1	-1	0
27,1 75,1	1,T	-1,T	-1,T	1,T	0

V

	1	2	3	4
1	1	1	1	1
49	1	1	-1	-1
28	1	-1	J	-J
76	1	-1	-J	J

VB

	1	2	3	4
1	1	1	1	1
49	1	1	-1	-1
28	1	-1	J	-J
76	1	-1	-J	J

SN

	1	2	3	4	5
1	1	1	1	1	2
49	1	1	1	1	-2
2,1 50,1	1,T	1,T	-1,T	-1,T	0
28 76	1	-1	1	-1	0
27,1 75,1	1,T	-1,T	-1,T	1,T	0

CA

	1	2	3	4	5
1	1	1	1	1	2
49	1	1	1	1	-2
2,1 50,1	1,T	1,T	-1,T	-1,T	0
28 76	1	-1	1	-1	0
27,1 75,1	1,T	-1,T	-1,T	1,T	0

GM

	1	2	3	4	5
1	1	1	1	1	2
49	1	1	1	1	-2
3,1 51,1	1	1	-1	-1	0
26 74	1	-1	1	-1	0
28,1 76,1	1	-1	-1	1	0

DT

	1	2	3	4	5
1	1	1	1	1	2
49	1	1	1	1	-2
3,1 51,1	1,T	1,T	-1,T	-1,T	0
26 74	1	-1	1	-1	0
28,1 76,1	1,T	-1,T	-1,T	1,T	0

SM

	1	2	3	4
1	1	1	1	1
49	1	1	-1	-1
76,1	1	-1	J	-J
28,1	1	-1	-J	J

X

	1	2	3	4	5
1	1	1	1	1	2
49	1	1	1	1	-2
3,1 51,1	1	1	-1	-1	0
26 74	1	-1	1	-1	0
28,1 76,1	1	-1	-1	1	0

Y

	1	2	3	4	5
1	1	1	1	1	2
49	1	1	1	1	-2
3,1 51,1	J	J	-J	-J	0
26 74	1	-1	1	-1	0
28,1 76,1	J	-J	-J	J	0

S

	1	2	3	4	5
1	1	1	1	1	2
49	1	1	1	1	-2
3,1 51,1	J	J	-J	-J	0
26 74	1	-1	1	-1	0
28,1 76,1	J	-J	-J	J	0

D

	1	2	3	4	5
1	1	1	1	1	2
49	1	1	1	1	-2
3,1 51,1	1,T	1,T	-1,T	-1,T	0
26 74	1	-1	1	-1	0
28,1 76,1	1,T	-1,T	-1,T	1,T	0

C

	1	2	3	4
1	1	1	1	1
49	1	1	-1	-1
76,1	J	-J	1	-1
28,1	J	-J	-1	1

V

	1	2	3	4
1	1	1	1	1
49	1	1	-1	-1
28,1	1,T	-1,T	-1,TJ	1,TJ
76,1	1,T	-1,T	1,TJ	-1,TJ

VA

	1	2	3	4
1	1	1	1	1
49	1	1	-1	-1
28,1	1,T	-1,T	1,TJ	-1,TJ
76,1	1,T	-1,T	-1,TJ	1,TJ

DA

	1	2	3	4	5
1	1	1	1	1	2
49	1	1	1	1	-2
3,1 51,1	1,T	1,T	-1,T	-1,T	0
26 74	1	-1	1	-1	0
28,1 76,1	1,T	-1,T	-1,T	1,T	0

TA

	1	2	3	4	5
1	1	1	1	1	2
49	1	1	1	1	-2
3,1 51,1	1,T	1,T	-1,T	-1,T	0
26 74	1	-1	1	-1	0
28,1 76,1	1,T	-1,T	-1,T	1,T	0

PBB2

GM

	1	2	3	4	5
1	1	1	1	1	2
49	1	1	1	1	-2
3 51	1	1	-1	-1	0
26,1 74,1	1	-1	1	-1	0
28,1 76,1	1	-1	-1	1	0

DT

	1	2	3	4	5
1	1	1	1	1	2
49	1	1	1	1	-2
3 51	1	1	-1	-1	0
26,1 74,1	1,T	-1,T	1,T	-1,T	0
28,1 76,1	1,T	-1,T	-1,T	1,T	0

SM

	1	2	3	4
1	1	1	1	1
49	1	1	-1	-1
76,1	1	-1	J	-J
28,1	1	-1	-J	J

X

	1	2	3	4	5
1	1	1	1	1	2
49	1	1	1	1	-2
3 51	1	1	-1	-1	0
26,1 74,1	1	-1	1	-1	0
28,1 76,1	1	-1	-1	1	0

Y

	1	2	3	4	5
1	1	1	1	1	2
49	1	1	1	1	-2
3 51	1	1	-1	-1	0
26,1 74,1	J	-J	J	-J	0
28,1 76,1	J	-J	-J	J	0

S

	1	2	3	4	5
1	1	1	1	1	2
49	1	1	1	1	-2
3 51	1	1	-1	-1	0
26,1 74,1	J	-J	J	-J	0
28,1 76,1	J	-J	-J	J	0

D

	1	2	3	4	5
1	1	1	1	1	2
49	1	1	1	1	-2
3 51	1	1	-1	-1	0
26,1 74,1	1,T	-1,T	1,T	-1,T	0
28,1 76,1	1,T	-1,T	-1,T	1,T	0

C

	1	2	3	4
1	1	1	1	1
49	1	1	-1	-1
76,1	J	-J	1	-1
28,1	J	-J	-1	1

V

	1	2	3	4
1	1	1	1	1
49	1	1	-1	-1
28,1	1,T	-1,T	-1,TJ	1,TJ
76,1	1,T	-1,T	1,TJ	-1,TJ

VA

	1	2	3	4
1	1	1	1	1
49	1	1	-1	-1
28,1	1,T	-1,T	1,TJ	-1,TJ
76,1	1,T	-1,T	-1,TJ	1,TJ

DA

	1	2	3	4	5
1	1	1	1	1	2
49	1	1	1	1	-2
3 51	1	1	-1	-1	0
26,1 74,1	1,T	-1,T	1,T	-1,T	0
28,1 76,1	1,T	-1,T	-1,T	1,T	0

TA

	1	2	3	4	5
1	1	1	1	1	2
49	1	1	1	1	-2
3 51	1	1	-1	-1	0
26,1 74,1	1,T	-1,T	1,T	-1,T	0
28,1 76,1	1,T	-1,T	-1,T	1,T	0

GROUP 28   P2MA

GM

GM	1	2	3	4	5
1	1	1	1	1	2
49	1	1	1	1	-2
4 52	1	1	-1	-1	0
26,1 74,1	1	-1	1	-1	0
27,1 75,1	1	-1	-1	1	0

DT

DT	1	2	3	4
1	1	1	1	1
49	1	1	-1	-1
26,1	1	-1	J	-J
74,1	1	-1	-J	J

SM

SM	1	2	3	4
1	1	1	1	1
49	1	1	-1	-1
75,1	1,T	-1,T	-1,TJ	1,TJ
27,1	1,T	-1,T	1,TJ	-1,TJ

X

X	1	2	3	4	5
1	2	1	1	1	1
49	2	-1	-1	-1	-1
4	0	J	J	-J	-J
26,1	0	J	-J	J	-J
75,1	0	-1	1	1	-1
52	0	-J	-J	J	J
74,1	0	-J	J	-J	J
27,1	0	1	-1	-1	1

Y

Y	1	2	3	4	5
1	1	1	1	1	2
49	1	1	1	1	-2
4 52	1	1	-1	-1	0
26,1 74,1	1	-1	1	-1	0
27,1 75,1	1	-1	-1	1	0

S

S	1	2	3	4	5
1	2	1	1	1	1
49	2	-1	-1	-1	-1
4	0	J	J	-J	-J
26,1	0	J	-J	J	-J
75,1	0	-1	1	1	-1
52	0	-J	-J	J	J
74,1	0	-J	J	-J	J
27,1	0	1	-1	-1	1

D

D	1	2	3	4
1	1	1	1	1
49	1	1	-1	-1
26,1	1	-1	J	-J
74,1	1	-1	-J	J

C

C	1	2	3	4
1	1	1	1	1
49	1	1	-1	-1
75,1	1,T	-1,T	-1,TJ	1,TJ
27,1	1,T	-1,T	1,TJ	-1,TJ

GROUP 29

GM

	1	2	3	4	5
1	1	1	1	1	2
49	1	1	1	1	-2
3 51	1	1	-1	-1	0
26,1 74,1	1	-1	1	-1	0
28,1 76,1	1	-1	-1	1	0

SM

	1	2	3	4
1	1	1	1	1
49	1	1	-1	-1
76,1	1,T	-1,T	-1,TJ	1,TJ
28,1	1,T	-1,T	1,TJ	-1,TJ

Y

	1	2	3	4	5
1	1	1	1	1	2
49	1	1	1	1	-2
3 51	1	1	-1	-1	0
26,1 74,1	1	-1	1	-1	0
28,1 76,1	1	-1	-1	1	0

S

	1	2	3	4	5
1	2	1	1	1	1
49	2	-1	-1	-1	-1
3	0	J	J	-J	-J
26,1	0	J	-J	J	-J
76,1	0	-1	1	1	-1
51	0	-J	-J	J	J
74,1	0	-J	J	-J	J
28,1	0	1	-1	-1	1

C

	1	2	3	4
1	1	1	1	1
49	1	1	-1	-1
76,1	1,T	-1,T	-1,TJ	1,TJ
28,1	1,T	-1,T	1,TJ	-1,TJ

VA

	1	2	3	4
1	1	1	1	1
49	1	1	-1	-1
28,1	1,T	-1,T	1,TJ	-1,TJ
76,1	1,T	-1,T	-1,TJ	1,TJ

TA

	1	2	3	4	5
1	1	1	1	1	2
49	1	1	1	1	-2
3 51	1	1	-1	-1	0
26,1 74,1	1	-1	1	-1	0
28,1 76,1	1	-1	-1	1	0

PAM2

DT

	1	2	3	4	5
1	1	1	1	1	2
49	1	1	1	1	-2
3 51	1	1	-1	-1	0
26,1 74,1	1	-1	1	-1	0
28,1 76,1	1	-1	-1	1	0

X

	1	2	3	4	5
1	2	1	1	1	1
49	2	-1	-1	-1	-1
3	0	J	J	-J	-J
26,1	0	J	-J	J	-J
76,1	0	-1	1	1	-1
51	0	-J	-J	J	J
74,1	0	-J	J	-J	J
28,1	0	1	-1	-1	1

D

	1	2	3	4	5
1	2	1	1	1	1
49	2	-1	-1	-1	-1
3	0	J	J	-J	-J
26,1	0	J	-J	J	-J
76,1	0	-1	1	1	-1
51	0	-J	-J	J	J
74,1	0	-J	J	-J	J
28,1	0	1	-1	-1	1

V

	1	2	3	4
1	1	1	1	1
49	1	1	-1	-1
28,1	1,T	-1,T	-1,TJ	1,TJ
76,1	1,T	-1,T	1,TJ	-1,TJ

DA

	1	2	3	4	5
1	2	1	1	1	1
49	2	-1	-1	-1	-1
3	0	J	J	-J	-J
26,1	0	-J	J	-J	J
28,1	0	-1	1	1	-1
51	0	-J	-J	J	J
74,1	0	J	-J	J	-J
76,1	0	1	-1	-1	1

GM

	1	2	3	4	5
1	1	1	1	1	2
49	1	1	1	1	-2
3,1 51,1	1	1	-1	-1	0
26,2 74,2	1	-1	1	-1	0
28,3 76,3	1	-1	-1	1	0

DT

	1	2	3	4	5
1	1	1	1	1	2
49	1	1	1	1	-2
3,1 51,1	1,T	1,T	-1,T	-1,T	0
26,2 74,2	1,T	-1,T	1,T	-1,T	0
28,3 76,3	1	-1	-1	1	0

SM

	1	2	3	4
1	1	1	1	1
49	1	1	-1	-1
76,3	1,T	-1,T	-1,TJ	1,TJ
28,3	1,T	-1,T	1,TJ	-1,TJ

Y

	1	2	3	4	5
1	1	1	1	1	2
49	1	1	1	1	-2
3,1 51,1	J	J	-J	-J	0
26,2 74,2	J	-J	J	-J	0
28,3 76,3	1	-1	-1	1	0

X

	1	2	3	4	5
1	2	1	1	1	1
49	2	-1	-1	-1	-1
3,1	0	J	J	-J	-J
26,2	0	J	-J	J	-J
76,3	0	-1	1	1	-1
51,1	0	-J	-J	J	J
74,2	0	-J	J	-J	J
28,3	0	1	-1	-1	1

S

	1	2	3	4	5
1	2	1	1	1	1
49	2	-1	-1	-1	-1
3,1	0	1	1	-1	-1
26,2	0	1	-1	1	-1
76,3	0	-1	1	1	-1
51,1	0	-1	-1	1	1
74,2	0	-1	1	-1	1
28,3	0	1	-1	-1	1

D

	1	2	3	4	5
1	2	1	1	1	1
49	2	-1	-1	-1	-1
3,1	0	-1,TJ	-1,TJ	1,TJ	1,TJ
26,2	0	-1,T	1,T	-1,T	1,T
76,3	0	-1	1	1	-1
51,1	0	1,TJ	1,TJ	-1,TJ	-1,TJ
74,2	0	1,T	-1,T	1,T	-1,T
28,3	0	1	-1	-1	1

C

	1	2	3	4
1	1	1	1	1
49	1	1	-1	-1
76,3	1,T	-1,T	-1,TJ	1,TJ
28,3	1,T	-1,T	1,TJ	-1,TJ

V

	1	2	3	4
1	1	1	1	1
49	1	1	-1	-1
28,3	1,T	-1,T	-1,TJ	1,TJ
76,3	1,T	-1,T	1,TJ	-1,TJ

VA

	1	2	3	4
1	1	1	1	1
49	1	1	-1	-1
28,3	1,T	-1,T	1,TJ	-1,TJ
76,3	1,T	-1,T	-1,TJ	1,TJ

DA

	1	2	3	4	5
1	2	1	1	1	1
49	2	-1	-1	-1	-1
3,1	0	-1,TJ	-1,TJ	1,TJ	1,TJ
26,2	0	1,T	-1,T	1,T	-1,T
28,3	0	-1	1	1	-1
51,1	0	1,TJ	1,TJ	-1,TJ	-1,TJ
74,2	0	-1,T	1,T	-1,T	1,T
76,3	0	1	-1	-1	1

TA

	1	2	3	4	5
1	1	1	1	1	2
49	1	1	1	1	-2
3,1 51,1	1,T	1,T	-1,T	-1,T	0
26,2 74,2	1,T	-1,T	1,T	-1,T	0
28,3 76,3	1	-1	-1	1	0

PNB2

GM

	1	2	3	4	5
1	1	1	1	1	2
49	1	1	1	1	-2
3 51	1	1	-1	-1	0
28,1 76,1	1	-1	1	-1	0
26,1 74,1	1	-1	-1	1	0

SM

	1	2	3	4
1	1	1	1	1
49	1	1	-1	-1
28,1	1,T	-1,T	-1,TJ	1,TJ
76,1	1,T	-1,T	1,TJ	-1,TJ

V

	1	2	3	4	5
1	1	1	1	1	2
49	1	1	1	1	-2
3 51	1	1	-1	-1	0
28,1 76,1	J	-J	J	-J	0
26,1 74,1	J	-J	-J	J	0

S

	1	2	3	4	5
1	2	1	1	1	1
49	2	-1	-1	-1	-1
3	0	J	J	-J	-J
28,1	0	J	-J	J	-J
74,1	0	-1	1	1	-1
51	0	-J	-J	J	J
76,1	0	-J	J	-J	J
26,1	0	1	-1	-1	1

C

	1	2	3	4
1	1	1	1	1
49	1	1	-1	-1
28,1	-1,T	1,T	-1,TJ	1,TJ
76,1	-1,T	1,T	1,TJ	-1,TJ

VA

	1	2	3	4
1	1	1	1	1
49	1	1	-1	-1
28,1	1,T	-1,T	1,TJ	-1,TJ
76,1	1,T	-1,T	-1,TJ	1,TJ

TA

	1	2	3	4	5
1	1	1	1	1	2
49	1	1	1	1	-2
3 51	1	1	-1	-1	0
28,1 76,1	1,T	-1,T	1,T	-1,T	0
26,1 74,1	1,T	-1,T	-1,T	1,T	0

DT

	1	2	3	4	5
1	1	1	1	1	2
49	1	1	1	1	-2
3 51	1	1	-1	-1	0
28,1 76,1	1,T	-1,T	1,T	-1,T	0
26,1 74,1	1,T	-1,T	-1,T	1,T	0

X

	1	2	3	4	5
1	2	1	1	1	1
49	2	-1	-1	-1	-1
3	0	J	J	-J	-J
28,1	0	1	-1	1	-1
74,1	0	J	-J	-J	J
51	0	-J	-J	J	J
76,1	0	-1	1	-1	1
26,1	0	-J	J	J	-J

D

	1	2	3	4	5
1	2	1	1	1	1
49	2	-1	-1	-1	-1
3	0	J	J	-J	-J
28,1	0	-1,TJ	1,TJ	-1,TJ	1,TJ
74,1	0	1,T	-1,T	-1,T	1,T
51	0	-J	-J	J	J
76,1	0	1,TJ	-1,TJ	1,TJ	-1,TJ
26,1	0	-1,T	1,T	1,T	-1,T

V

	1	2	3	4
1	1	1	1	1
49	1	1	-1	-1
28,1	1,T	-1,T	-1,TJ	1,TJ
76,1	1,T	-1,T	1,TJ	-1,TJ

DA

	1	2	3	4	5
1	2	1	1	1	1
49	2	-1	-1	-1	-1
3	0	J	J	-J	-J
28,1	0	1,TJ	-1,TJ	1,TJ	-1,TJ
26,1	0	1,T	-1,T	-1,T	1,T
51	0	-J	-J	J	J
76,1	0	-1,TJ	1,TJ	-1,TJ	1,TJ
74,1	0	-1,T	1,T	1,T	-1,T

PNM2(1)

GM

	1	2	3	4	5
1	1	1	1	1	2
49	1	1	1	1	-2
3,1 51,1	1	1	-1	-1	0
26 74	1	-1	1	-1	0
28,1 76,1	1	-1	-1	1	0

DT

	1	2	3	4	5
1	1	1	1	1	2
49	1	1	1	1	-2
3,1 51,1	1,T	1,T	-1,T	-1,T	0
26 74	1	-1	1	-1	0
28,1 76,1	1,T	-1,T	-1,T	1,T	0

SM

	1	2	3	4
1	1	1	1	1
49	1	1	-1	-1
76,1	1,T	-1,T	-1,TJ	1,TJ
28,1	1,T	-1,T	1,TJ	-1,TJ

X

	1	2	3	4	5
1	2	1	1	1	1
49	2	-1	-1	-1	-1
3,1	0	J	J	-J	-J
26	0	J	-J	J	-J
76,1	0	1	-1	-1	1
51,1	0	-J	-J	J	J
74	0	-J	J	-J	J
28,1	0	-1	1	1	-1

Y

	1	2	3	4	5
1	1	1	1	1	2
49	1	1	1	1	-2
3,1 51,1	J	J	-J	-J	0
26 74	1	-1	1	-1	0
28,1 76,1	J	-J	-J	J	0

S

	1	2	3	4	5
1	2	1	1	1	1
49	2	-1	-1	-1	-1
3,1	0	1	1	-1	-1
26	0	J	-J	J	-J
76,1	0	-J	J	J	-J
51,1	0	-1	-1	1	1
74	0	-J	J	-J	J
28,1	0	J	-J	-J	J

D

	1	2	3	4	5
1	2	1	1	1	1
49	2	-1	-1	-1	-1
3,1	0	-1,T	-1,T	1,T	1,T
26	0	J	-J	J	-J
76,1	0	1,TJ	-1,TJ	-1,TJ	1,TJ
51,1	0	1,T	1,T	-1,T	-1,T
74	0	-J	J	-J	J
28,1	0	-1,TJ	1,TJ	1,TJ	-1,TJ

C

	1	2	3	4
1	1	1	1	1
49	1	1	-1	-1
76,1	-1,T	1,T	-1,TJ	1,TJ
28,1	-1,T	1,T	1,TJ	-1,TJ

V

	1	2	3	4
1	1	1	1	1
49	1	1	-1	-1
28,1	1,T	-1,T	-1,TJ	1,TJ
76,1	1,T	-1,T	1,TJ	-1,TJ

VA

	1	2	3	4
1	1	1	1	1
49	1	1	-1	-1
28,1	1,T	-1,T	1,TJ	-1,TJ
76,1	1,T	-1,T	-1,TJ	1,TJ

DA

	1	2	3	4	5
1	2	1	1	1	1
49	2	-1	-1	-1	-1
3,1	0	-1,T	-1,T	1,T	1,T
26	0	-J	J	-J	J
28,1	0	1,TJ	-1,TJ	-1,TJ	1,TJ
51,1	0	1,T	1,T	-1,T	-1,T
74	0	J	-J	J	-J
76,1	0	-1,TJ	1,TJ	1,TJ	-1,TJ

TA

	1	2	3	4	5
1	1	1	1	1	2
49	1	1	1	1	-2
3,1 51,1	1,T	1,T	-1,T	-1,T	0
26 74	1	-1	1	-1	0
28,1 76,1	1,T	-1,T	-1,T	1,T	0

GROUP 33   P2BA

GM

	1	2	3	4	5
1	1	1	1	1	2
49	1	1	1	1	-2
4 52	1	1	-1	-1	0
26,1 74,1	1	-1	1	-1	0
27,1 75,1	1	-1	-1	1	0

DT

	1	2	3	4
1	1	1	1	1
49	1	1	-1	-1
26,1	1,T	-1,T	-1,TJ	1,TJ
74,1	1,T	-1,T	1,TJ	-1,TJ

SM

	1	2	3	4
1	1	1	1	1
49	1	1	-1	-1
75,1	1,T	-1,T	-1,TJ	1,TJ
27,1	1,T	-1,T	1,TJ	-1,TJ

X

	1	2	3	4	5
1	2	1	1	1	1
49	2	-1	-1	-1	-1
4	0	J	J	-J	-J
26,1	0	J	-J	J	-J
75,1	0	-1	1	1	-1
52	0	-J	-J	J	J
74,1	0	-J	J	-J	J
27,1	0	1	-1	-1	1

Y

	1	2	3	4	5
1	2	1	1	1	1
49	2	-1	-1	-1	-1
4	0	J	J	-J	-J
26,1	0	1	-1	1	-1
75,1	0	J	-J	-J	J
52	0	-J	-J	J	J
74,1	0	-1	1	-1	1
27,1	0	-J	J	J	-J

S

	1	2	3	4	5
1	1	1	1	1	2
49	1	1	1	1	-2
4 52	1	1	-1	-1	0
26,1 74,1	J	-J	J	-J	0
27,1 75,1	J	-J	-J	J	0

C

	1	2	3	4
1	1	1	1	1
49	1	1	-1	-1
75,1	-1,TJ	1,TJ	-1,T	1,T
27,1	-1,TJ	1,TJ	1,T	-1,T

D

	1	2	3	4
1	1	1	1	1
49	1	1	-1	-1
26,1	-1,TJ	1,TJ	-1,T	1,T
74,1	-1,TJ	1,TJ	1,T	-1,T

GROUP 34   C2MM

GM

	1	2	3	4	5
1	1	1	1	1	2
49	1	1	1	1	-2
4 52	1	1	-1	-1	0
26 74	1	-1	1	-1	0
27 75	1	-1	-1	1	0

DT

	1	2	3	4
1	1	1	1	1
49	1	1	-1	-1
26	1	-1	J	-J
74	1	-1	-J	J

SM

	1	2	3	4
1	1	1	1	1
49	1	1	-1	-1
75	1	-1	J	-J
27	1	-1	-J	J

YA

	1	2	3	4	5
1	1	1	1	1	2
49	1	1	1	1	-2
4 52	1	1	-1	-1	0
26 74	1	-1	1	-1	0
27 75	1	-1	-1	1	0

YB

	1	2	3	4	5
1	1	1	1	1	2
49	1	1	1	1	-2
4 52	1	1	-1	-1	0
26 74	1	-1	1	-1	0
27 75	1	-1	-1	1	0

S

	1	2	3	4
1	1	1	1	1
49	1	1	-1	-1
4	1	-1	J	-J
52	1	-1	-J	J

C

	1	2	3	4
1	1	1	1	1
49	1	1	-1	-1
75	1	-1	J	-J
27	1	-1	-J	J

F

	1	2	3	4
1	1	1	1	1
49	1	1	-1	-1
26	1	-1	J	-J
74	1	-1	-J	J

GROUP 35   CMM2

GM

	1	2	3	4	5
1	1	1	1	1	2
49	1	1	1	1	-2
3 51	1	1	-1	-1	0
26 74	1	-1	1	-1	0
28 76	1	-1	-1	1	0

DT

	1	2	3	4	5
1	1	1	1	1	2
49	1	1	1	1	-2
3 51	1	1	-1	-1	0
26 74	1	-1	1	-1	0
28 76	1	-1	-1	1	0

SM

	1	2	3	4
1	1	1	1	1
49	1	1	-1	-1
28	1	-1	J	-J
76	1	-1	-J	J

YA

	1	2	3	4	5
1	1	1	1	1	2
49	1	1	1	1	-2
3 51	1	1	-1	-1	0
26 74	1	-1	1	-1	0
28 76	1	-1	-1	1	0

YB

	1	2	3	4	5
1	1	1	1	1	2
49	1	1	1	1	-2
3 51	1	1	-1	-1	0
26 74	1	-1	1	-1	0
28 76	1	-1	-1	1	0

S

	1	2	3	4
1	1	1	1	1
49	1	1	-1	-1
28	1	-1	J	-J
76	1	-1	-J	J

C

	1	2	3	4	5
1	1	1	1	1	2
49	1	1	1	1	-2
3 51	1	1	-1	-1	0
26 74	1	-1	1	-1	0
28 76	1	-1	-1	1	0

F

	1	2	3	4	5
1	1	1	1	1	2
49	1	1	1	1	-2
3 51	1	1	-1	-1	0
26 74	1	-1	1	-1	0
28 76	1	-1	-1	1	0

P

	1	2	3	4
1	1	1	1	1
49	1	1	-1	-1
28	1	-1	J	-J
76	1	-1	-J	J

DA

	1	2	3	4	5
1	1	1	1	1	2
49	1	1	1	1	-2
3 51	1	1	-1	-1	0
26 74	1	-1	1	-1	0
28 76	1	-1	-1	1	0

FA

	1	2	3	4	5
1	1	1	1	1	2
49	1	1	1	1	-2
3 51	1	1	-1	-1	0
26 74	1	-1	1	-1	0
28 76	1	-1	-1	1	0

PA

	1	2	3	4
1	1	1	1	1
49	1	1	-1	-1
28	1	-1	-J	J
76	1	-1	J	-J

GROUP 36   CAM2

GM

	1	2	3	4	5
1	1	1	1	1	2
49	1	1	1	1	-2
3 51	1	1	-1	-1	0
26,1 74,1	1	-1	1	-1	0
28,1 76,1	1	-1	-1	1	0

DT

	1	2	3	4	5
1	1	1	1	1	2
49	1	1	1	1	-2
3 51	1	1	-1	-1	0
26,1 74,1	1,T	-1,T	1,T	-1,T	0
28,1 76,1	1,T	-1,T	-1,T	1,T	0

SM

	1	2	3	4
1	1	1	1	1
49	1	1	-1	-1
28,1	1	-1	J	-J
76,1	1	-1	-J	J

YA

	1	2	3	4	5
1	1	1	1	1	2
49	1	1	1	1	-2
3 51	1	1	-1	-1	0
26,1 74,1	1	-1	1	-1	0
28,1 76,1	1	-1	-1	1	0

YB

	1	2	3	4	5
1	1	1	1	1	2
49	1	1	1	1	-2
3 51	1	1	-1	-1	0
26,1 74,1	1	-1	1	-1	0
28,1 76,1	1	-1	-1	1	0

S

	1	2	3	4
1	1	1	1	1
49	1	1	-1	-1
28,1	J	-J	1	-1
76,1	J	-J	-1	1

C

	1	2	3	4
1	1	1	1	1
49	1	1	-1	-1
28,1	1	-1	J	-J
76,1	1	-1	-J	J

F

	1	2	3	4	5
1	1	1	1	1	2
49	1	1	1	1	-2
3 51	1	1	-1	-1	0
26,1 74,1	1,T	-1,T	1,T	-1,T	0
28,1 76,1	1,T	-1,T	-1,T	1,T	0

P

	1	2	3	4
1	1	1	1	1
49	1	1	-1	-1
28,1	1,T	-1,T	-1,TJ	1,TJ
76,1	1,T	-1,T	1,TJ	-1,TJ

DA

	1	2	3	4	5
1	1	1	1	1	2
49	1	1	1	1	-2
3 51	1	1	-1	-1	0
26,1 74,1	1,T	-1,T	1,T	-1,T	0
28,1 76,1	1,T	-1,T	-1,T	1,T	0

FA

	1	2	3	4	5
1	1	1	1	1	2
49	1	1	1	1	-2
3 51	1	1	-1	-1	0
26,1 74,1	1,T	-1,T	1,T	-1,T	0
28,1 76,1	1,T	-1,T	-1,T	1,T	0

PA

	1	2	3	4
1	1	1	1	1
49	1	1	-1	-1
28,1	1,T	-1,T	1,TJ	-1,TJ
76,1	1,T	-1,T	-1,TJ	1,TJ

GM

	1+	2+	3+	4+	1-	2-	3-	4-	5+	5-
1	1	1	1	1	1	1	1	1	2	2
49	1	1	1	1	1	1	1	1	-2	-2
4 52	1	1	-1	-1	1	1	-1	-1	0	0
2 50	1	-1	1	-1	1	-1	1	-1	0	0
3 51	1	-1	-1	1	1	-1	-1	1	0	0
26 74	1	-1	1	-1	-1	1	-1	1	0	0
27 75	1	-1	-1	1	-1	1	1	-1	0	0
73	1	1	1	1	-1	-1	-1	-1	-2	2
28 76	1	1	-1	-1	-1	-1	1	1	0	0
25	1	1	1	1	-1	-1	-1	-1	2	-2

DT

	1	2	3	4	5
1	1	1	1	1	2
49	1	1	1	1	-2
3 51	1	1	-1	-1	0
26 74	1	-1	1	-1	0
28 76	1	-1	-1	1	0

SM

	1	2	3	4	5
1	1	1	1	1	2
49	1	1	1	1	-2
2 50	1	1	-1	-1	0
27 75	1	-1	1	-1	0
28 76	1	-1	-1	1	0

X

	1+	2+	3+	4+	1-	2-	3-	4-	5+	5-
1	1	1	1	1	1	1	1	1	2	2
49	1	1	1	1	1	1	1	1	-2	-2
4 52	1	1	-1	-1	1	1	-1	-1	0	0
2 50	1	-1	1	-1	1	-1	1	-1	0	0
3 51	1	-1	-1	1	1	-1	-1	1	0	0
26 74	1	-1	1	-1	-1	1	-1	1	0	0
27 75	1	-1	-1	1	-1	1	1	-1	0	0
73	1	1	1	1	-1	-1	-1	-1	-2	2
28 76	1	1	-1	-1	-1	-1	1	1	0	0
25	1	1	1	1	-1	-1	-1	-1	2	-2

Y

	1+	2+	3+	4+	1-	2-	3-	4-	5+	5-
1	1	1	1	1	1	1	1	1	2	2
49	1	1	1	1	1	1	1	1	-2	-2
4 52	1	1	-1	-1	1	1	-1	-1	0	0
2 50	1	-1	1	-1	1	-1	1	-1	0	0
3 51	1	-1	-1	1	1	-1	-1	1	0	0
26 74	1	-1	1	-1	-1	1	-1	1	0	0
27 75	1	-1	-1	1	-1	1	1	-1	0	0
73	1	1	1	1	-1	-1	-1	-1	-2	2
28 76	1	1	-1	-1	-1	-1	1	1	0	0
25	1	1	1	1	-1	-1	-1	-1	2	-2

S	1+	2+	3+	4+	1-	2-	3-	4-	5+	5-
1	1	1	1	1	1	1	1	1	2	2
49	1	1	1	1	1	1	1	1	-2	-2
4 52	1	1	-1	-1	1	1	-1	-1	0	0
2 50	1	-1	1	-1	1	-1	1	-1	0	0
3 51	1	-1	-1	1	1	-1	-1	1	0	0
26 74	1	-1	1	-1	-1	1	-1	1	0	0
27 75	1	-1	-1	1	-1	1	1	-1	0	0
73	1	1	1	1	-1	-1	-1	-1	-2	2
28 76	1	1	-1	-1	-1	-1	1	1	0	0
25	1	1	1	1	-1	-1	-1	-1	2	-2

D	1	2	3	4	5
1	1	1	1	1	2
49	1	1	1	1	-2
3 51	1	1	-1	-1	0
26 74	1	-1	1	-1	0
28 76	1	-1	-1	1	0

C	1	2	3	4	5
1	1	1	1	1	2
49	1	1	1	1	-2
2 50	1	1	-1	-1	0
27 75	1	-1	1	-1	0
28 76	1	-1	-1	1	0

V	1	2	3	4
1	1	1	1	1
49	1	1	-1	-1
28	1	-1	J	-J
76	1	-1	-J	J

P2/A2/M2/A

GM

	1+	2+	3+	4+	1-	2-	3-	4-	5+	5-
1	1	1	1	1	1	1	1	1	2	2
49	1	1	1	1	1	1	1	1	-2	-2
2 50	1	1	-1	-1	1	1	-1	-1	0	0
4,1 52,1	1	-1	1	-1	1	-1	1	-1	0	0
3,1 51,1	1	-1	-1	1	1	-1	-1	1	0	0
28,1 76,1	1	-1	1	-1	-1	1	-1	1	0	0
27,1 75,1	1	-1	-1	1	-1	1	1	-1	0	0
73	1	1	1	1	-1	-1	-1	-1	-2	2
26 74	1	1	-1	-1	-1	-1	1	1	0	0
25	1	1	1	1	-1	-1	-1	-1	2	-2

DT

	1	2	3	4	5
1	1	1	1	1	2
49	1	1	1	1	-2
3,1 51,1	1	1	-1	-1	0
28,1 76,1	1	-1	1	-1	0
26 74	1	-1	-1	1	0

SM

	1	2	3	4	5
1	1	1	1	1	2
49	1	1	1	1	-2
2 50	1	1	-1	-1	0
28,1 76,1	1,T	-1,T	1,T	-1,T	0
27,1 75,1	1,T	-1,T	-1,T	1,T	0

X

	1	2	3	4
1	2	2	2	2
49	2	2	-2	-2
2 50	2	-2	0	0
3,1 4,1 25				
27,1 28,1 51,1				
52,1 73 75,1				
76,1	0	0	0	0
74	0	0	-2J	2J
26	0	0	2J	-2J

Y

	1+	2+	3+	4+	1-	2-	3-	4-	5+	5-
1	1	1	1	1	1	1	1	1	2	2
49	1	1	1	1	1	1	1	1	-2	-2
2 50	1	1	-1	-1	1	1	-1	-1	0	0
4,1 52,1	1	-1	1	-1	1	-1	1	-1	0	0
3,1 51,1	1	-1	-1	1	1	-1	-1	1	0	0
28,1 76,1	1	-1	1	-1	-1	1	-1	1	0	0
27,1 75,1	1	-1	-1	1	-1	1	1	-1	0	0
73	1	1	1	1	-1	-1	-1	-1	-2	2
26 74	1	1	-1	-1	-1	-1	1	1	0	0
25	1	1	1	1	-1	-1	-1	-1	2	-2

S

	1	2	3	4
1	2	2	2	2
49	2	2	-2	-2
2 50	2	-2	0	0
3,1 4,1 25				
27,1 28,1 51,1				
52,1 73 75,1				
76,1	0	0	0	0
74	0	0	-2J	2J
26	0	0	2J	-2J

D

	1	2	3	4	5
1	2	1	1	1	1
49	2	-1	-1	-1	-1
51,1	0	J	J	-J	-J
28,1	0	1	-1	1	-1
74	0	-J	J	J	-J
3,1	0	-J	-J	J	J
76,1	0	-1	1	-1	1
26	0	J	-J	-J	J

C

	1	2	3	4	5
1	1	1	1	1	2
49	1	1	1	1	-2
2 50	1	1	-1	-1	0
28,1 76,1	1,T	-1,T	1,T	-1,T	0
27,1 75,1	1,T	-1,T	-1,T	1,T	0

V

	1	2	3	4
1	1	1	1	1
49	1	1	-1	-1
28,1	1,T	-1,T	-1,TJ	1,TJ
76,1	1,T	-1,T	1,TJ	-1,TJ

GM

	1+	2+	3+	4+	1-	2-	3-	4-	5+	5-
1	1	1	1	1	1	1	1	1	2	2
49	1	1	1	1	1	1	1	1	-2	-2
4 52	1	1	-1	-1	1	1	-1	-1	0	0
2 50	1	-1	1	-1	1	-1	1	-1	0	0
3 51	1	-1	-1	1	1	-1	-1	1	0	0
26,1 74,1	1	-1	1	-1	-1	1	-1	1	0	0
27,1 75,1	1	-1	-1	1	-1	1	1	-1	0	0
73,1	1	1	1	1	-1	-1	-1	-1	-2	2
28,1 76,1	1	1	-1	-1	-1	-1	1	1	0	0
25,1	1	1	1	1	-1	-1	-1	-1	2	-2

DT

	1	2	3	4	5
1	1	1	1	1	2
49	1	1	1	1	-2
3 51	1	1	-1	-1	0
26,1 74,1	1,T	-1,T	1,T	-1,T	0
28,1 76,1	1,T	-1,T	-1,T	1,T	0

SM

	1	2	3	4	5
1	1	1	1	1	2
49	1	1	1	1	-2
2 50	1	1	-1	-1	0
27,1 75,1	1,T	-1,T	1,T	-1,T	0
28,1 76,1	1,T	-1,T	-1,T	1,T	0

X

	1	2	3	4
1	2	2	2	2
49	2	2	-2	-2
3 4 25,1				
27,1 28,1 51				
52 73,1 75,1				
76,1	0	0	0	0
26,1	0	0	2J	-2J
2 50	2	-2	0	0
74,1	0	0	-2J	2J

Y

	1	2	3	4
1	2	2	2	2
49	2	2	-2	-2
2 4 25,1				
26,1 28,1 50				
52 73,1 74,1				
76,1	0	0	0	0
75,1	0	0	2J	-2J
3 51	2	-2	0	0
27,1	0	0	-2J	2J

S

	1	2	3	4
1	2	2	2	2
49	2	2	-2	-2
4 52	2	-2	0	0
2 3 25,1				
26,1 27,1 50				
51 73,1 74,1				
75,1	0	0	0	0
76,1	0	0	2J	-2J
28,1	0	0	-2J	2J

D

	1	2	3	4	5
1	2	1	1	1	1
49	2	-1	-1	-1	-1
51	0	J	J	-J	-J
26,1	0	-1,T	1,T	-1,T	1,T
76,1	0	-1,TJ	1,TJ	1,TJ	-1,TJ
3	0	-J	-J	J	J
74,1	0	1,T	-1,T	1,T	-1,T
28,1	0	1,TJ	-1,TJ	-1,TJ	1,TJ

C

	1	2	3	4	5
1	2	1	1	1	1
49	2	-1	-1	-1	-1
2	0	J	J	-J	-J
75,1	0	-1,T	1,T	-1,T	1,T
28,1	0	-1,TJ	1,TJ	1,TJ	-1,TJ
50	0	-J	-J	J	J
27,1	0	1,T	-1,T	1,T	-1,T
76,1	0	1,TJ	-1,TJ	-1,TJ	1,TJ

V

	1	2	3	4
1	1	1	1	1
49	1	1	-1	-1
28,1	1,T	-1,T	-1,TJ	1,TJ
76,1	1,T	-1,T	1,TJ	-1,TJ

GM

	1+	2+	3+	4+	1-	2-	3-	4-	5+	5-
1	1	1	1	1	1	1	1	1	2	2
49	1	1	1	1	1	1	1	1	-2	-2
3,1 51,1	1	1	-1	-1	1	1	-1	-1	0	0
2,1 50,1	1	-1	1	-1	1	-1	1	-1	0	0
4 52	1	-1	-1	1	1	-1	-1	1	0	0
26,1 74,1	1	-1	1	-1	-1	1	1	-1	0	0
28 76	1	-1	-1	1	-1	1	1	-1	0	0
73	1	1	1	1	-1	-1	-1	-1	-2	2
27,1 75,1	1	1	-1	-1	-1	-1	1	1	0	0
25	1	1	1	1	-1	-1	-1	-1	2	-2

DT

	1	2	3	4	5
1	1	1	1	1	2
49	1	1	1	1	-2
3,1 51,1	1	1	-1	-1	0
26,1 74,1	1	-1	1	-1	0
28 76	1	-1	-1	1	0

SM

	1	2	3	4	5
1	1	1	1	1	2
49	1	1	1	1	-2
2,1 50,1	1,T	1,T	-1,T	-1,T	0
28 76	1	-1	1	-1	0
27,1 75,1	1,T	-1,T	-1,T	1,T	0

X

	1	2	3	4
1	2	2	2	2
49	2	2	-2	-2
2,1 3,1 25				
26,1 27,1 50,1				
51,1 73 74,1				
75,1	0	0	0	0
52	0	0	-2J	2J
28 76	2	-2	0	0
4	0	0	2J	-2J

Y

	1+	2+	3+	4+	1-	2-	3-	4-	5+	5-
1	1	1	1	1	1	1	1	1	2	2
49	1	1	1	1	1	1	1	1	-2	-2
3,1 51,1	1	1	-1	-1	1	1	-1	-1	0	0
2,1 50,1	1	-1	1	-1	1	-1	1	-1	0	0
4 52	1	-1	-1	1	1	-1	-1	1	0	0
26,1 74,1	1	-1	1	-1	-1	1	1	-1	0	0
28 76	1	-1	-1	1	-1	1	1	-1	0	0
73	1	1	1	1	-1	-1	-1	-1	-2	2
27,1 75,1	1	1	-1	-1	-1	-1	1	1	0	0
25	1	1	1	1	-1	-1	-1	-1	2	-2

S

	1	2	3	4
1	2	2	2	2
49	2	2	-2	-2
2,1 3,1 25 26,1 27,1 50,1 51,1 73 74,1 75,1	0	0	0	0
52	0	0	-2J	2J
28 76	2	-2	0	0
4	0	0	2J	-2J

D

	1	2	3	4	5
1	1	1	1	1	2
49	1	1	1	1	-2
3,1 51,1	1	1	-1	-1	0
26,1 74,1	1	-1	1	-1	0
28 76	1	-1	-1	1	0

C

	1	2	3	4	5
1	1	1	1	1	2
49	1	1	1	1	-2
2,1 50,1	1,T	1,T	-1,T	-1,T	0
28 76	1	-1	1	-1	0
27,1 75,1	1,T	-1,T	-1,T	1,T	0

V

	1	2	3	4
1	1	1	1	1
49	1	1	-1	-1
28	1	-1	J	-J
76	1	-1	-J	J

GM

	1+	2+	3+	4+	1-	2-	3-	4-	5+	5-
1	1	1	1	1	1	1	1	1	2	2
49	1	1	1	1	1	1	1	1	-2	-2
4,1 52,1	1	1	-1	-1	1	1	-1	-1	0	0
2,1 50,1	1	-1	1	-1	1	-1	1	-1	0	0
3 51	1	-1	-1	1	1	-1	-1	1	0	0
26,1 74,1	1	-1	1	-1	-1	1	-1	1	0	0
27 75	1	-1	-1	1	-1	1	1	-1	0	0
73	1	1	1	1	-1	-1	-1	-1	-2	2
28,1 76,1	1	1	-1	-1	-1	-1	1	1	0	0
25	1	1	1	1	-1	-1	-1	-1	2	-2

DT

	1	2	3	4	5
1	1	1	1	1	2
49	1	1	1	1	-2
3 51	1	1	-1	-1	0
26.1 74,1	1	-1	1	-1	0
28,1 76,1	1	-1	-1	1	0

SM

	1	2	3	4	5
1	1	1	1	1	2
49	1	1	1	1	-2
2,1 50,1	1,T	1,T	-1,T	-1,T	0
27 75	1	-1	1	-1	0
28,1 76,1	1,T	-1,T	-1,T	1,T	0

X

	1	2	3	4
1	2	2	2	2
49	2	2	-2	-2
2,1 4,1 25				
26,1 28,1 50,1				
52,1 73 74,1				
76,1	0	0	0	0
51	0	0	-2J	2J
27 75	2	-2	0	0
3	0	0	2J	-2J

Y

	1+	2+	3+	4+	1-	2-	3-	4-	5+	5-
1	1	1	1	1	1	1	1	1	2	2
49	1	1	1	1	1	1	1	1	-2	-2
4,1 52,1	1	1	-1	-1	1	1	-1	-1	0	0
2,1 50,1	1	-1	1	-1	1	-1	1	-1	0	0
3 51	1	-1	-1	1	1	-1	-1	1	0	0
26,1 74,1	1	-1	1	-1	-1	1	-1	1	0	0
27 75	1	-1	-1	1	-1	1	1	-1	0	0
73	1	1	1	1	-1	-1	-1	-1	-2	2
28,1 76,1	1	1	-1	-1	-1	-1	1	1	0	0
25	1	1	1	1	-1	-1	-1	-1	2	-2

S

	1	2	3	4
1	2	2	2	2
49	2	2	-2	-2
2,1 4,1 25				
26,1 28,1 50,1				
52,1 73 74,1				
76,1	0	0	0	0
51	0	0	-2J	2J
27 75	2	-2	0	0
3	0	0	2J	-2J

D

	1	2	3	4	5
1	2	1	1	1	1
49	2	-1	-1	-1	-1
51	0	J	J	-J	-J
26,1	0	J	-J	J	-J
76,1	0	-1	1	1	-1
3	0	-J	-J	J	J
74,1	0	-J	J	-J	J
28,1	0	1	-1	-1	1

C

	1	2	3	4	5
1	1	1	1	1	2
49	1	1	1	1	-2
2,1 50,1	1,T	1,T	-1,T	-1,T	0
27 75	1	-1	1	-1	0
28,1 76,1	1,T	-1,T	-1,T	1,T	0

V

	1	2	3	4
1	1	1	1	1
49	1	1	-1	-1
28,1	1,T	-1,T	-1,TJ	1,TJ
76,1	1,T	-1,T	1,TJ	-1,TJ

GM

	1+	2+	3+	4+	1−	2−	3−	4−	5+	5−
1	1	1	1	1	1	1	1	1	2	2
49	1	1	1	1	1	1	1	1	−2	−2
3,1 51,1	1	1	−1	−1	1	1	−1	−1	0	0
2 50	1	−1	1	−1	1	−1	1	−1	0	0
4,1 52,1	1	−1	−1	1	1	−1	−1	1	0	0
26 74	1	−1	1	−1	−1	1	−1	1	0	0
28,1 76,1	1	−1	−1	1	−1	1	1	−1	0	0
73	1	1	1	1	−1	−1	−1	−1	−2	2
27,1 75,1	1	1	−1	−1	−1	−1	1	1	0	0
25	1	1	1	1	−1	−1	−1	−1	2	−2

DT

	1	2	3	4	5
1	1	1	1	1	2
49	1	1	1	1	−2
3,1 51,1	1,T	1,T	−1,T	−1,T	0
26 74	1	−1	1	−1	0
28,1 76,1	1,T	−1,T	−1,T	1,T	0

SM

	1	2	3	4	5
1	1	1	1	1	2
49	1	1	1	1	−2
2 50	1	1	−1	−1	0
28,1 76,1	1,T	−1,T	1,T	−1,T	0
27,1 75,1	1,T	−1,T	−1,T	1,T	0

X

	1	2	3	4
1	2	2	2	2
49	2	2	−2	−2
3,1 4,1 25 27,1 28,1 51,1 52,1 73 75,1 76,1	0	0	0	0
26	0	0	2J	−2J
2 50	2	−2	0	0
74	0	0	−2J	2J

Y

	1	2	3	4
1	2	2	2	2
49	2	2	−2	−2
3,1 4,1 25 27,1 28,1 51,1 52,1 73 75,1 76,1	0	0	0	0
2	0	0	2J	−2J
26 74	2	−2	0	0
50	0	0	−2J	2J

S	1+	1-	2+	3+	4+	5+	2-	3-	4-	5-
1	2	2	1	1	1	1	1	1	1	1
49	2	2	-1	-1	-1	-1	-1	-1	-1	-1
3,1	0	0	1	1	-1	-1	1	1	-1	-1
2	0	0	J	-J	J	-J	J	-J	J	-J
52,1	0	0	-J	J	J	-J	-J	J	J	-J
26	0	0	J	-J	J	-J	-J	J	-J	J
76,1	0	0	-J	J	J	-J	J	-J	-J	J
75,1	0	0	-1	-1	1	1	1	1	-1	-1
73	2	-2	-1	-1	-1	-1	1	1	1	1
51,1	0	0	-1	-1	1	1	-1	-1	1	1
50	0	0	-J	J	-J	J	-J	J	-J	J
4,1	0	0	J	-J	-J	J	J	-J	-J	J
74	0	0	-J	J	-J	J	J	-J	J	-J
28,1	0	0	J	-J	-J	J	-J	J	J	-J
27,1	0	0	1	1	-1	-1	-1	-1	1	1
25	2	-2	1	1	1	1	-1	-1	-1	-1

D		1	2	3	4	5
1		2	1	1	1	1
49		2	-1	-1	-1	-1
3,1		0	-1,T	-1,T	1,T	1,T
26		0	J	-J	J	-J
76,1		0	1,TJ	-1,TJ	-1,TJ	1,TJ
51,1		0	1,T	1,T	-1,T	-1,T
74		0	-J	J	-J	J
28,1		0	-1,TJ	1,TJ	1,TJ	-1,TJ

C		1	2	3	4	5
1		2	1	1	1	1
49		2	-1	-1	-1	-1
2		0	J	J	-J	-J
76,1		0	-1,TJ	1,TJ	-1,TJ	1,TJ
27,1		0	1,T	-1,T	-1,T	1,T
50		0	-J	-J	J	J
28,1		0	1,TJ	-1,TJ	1,TJ	-1,TJ
75,1		0	-1,T	1,T	1,T	-1,T

V		1	2	3	4
1		1	1	1	1
49		1	1	-1	-1
28,1		1,T	-1,T	-1,TJ	1,TJ
76,1		1,T	-1,T	1,TJ	-1,TJ

GM

	1+	2+	3+	4+	1-	2-	3-	4-	5+	5-
1	1	1	1	1	1	1	1	1	2	2
49	1	1	1	1	1	1	1	1	-2	-2
2,1 50,1	1	1	-1	-1	1	1	-1	-1	0	0
3,2 51,2	1	-1	1	-1	1	-1	1	-1	0	0
4,3 52,3	1	-1	-1	1	1	-1	-1	1	0	0
27,2 75,2	1	-1	1	-1	-1	1	-1	1	0	0
28,3 76,3	1	-1	-1	1	-1	1	1	-1	0	0
73	1	1	1	1	-1	-1	-1	-1	-2	2
26,1 74,1	1	1	-1	-1	-1	-1	1	1	0	0
25	1	1	1	1	-1	-1	-1	-1	2	-2

DT

	1	2	3	4	5
1	1	1	1	1	2
49	1	1	1	1	-2
3,2 51,2	1,T	1,T	-1,T	-1,T	0
28,3 76,3	1	-1	1	-1	0
26,1 74,1	1,T	-1,T	-1,T	1,T	0

SM

	1	2	3	4	5
1	1	1	1	1	2
49	1	1	1	1	-2
2,1 50,1	1	1	-1	-1	0
27,2 75,2	1,T	-1,T	1,T	-1,T	0
28,3 76,3	1,T	-1,T	-1,T	1,T	0

X

	1	2	3	4
1	2	2	2	2
49	2	2	-2	-2
2,1 50,1	2	-2	0	0
3,2 4,3 25				
27,2 28,3 51,2				
52,3 73 75,2				
76,3	0	0	0	0
74,1	0	0	-2J	2J
26,1	0	0	2J	-2J

Y

	1	2	3	4
1	2	2	2	2
49	2	2	-2	-2
2,1 3,2 25				
26,1 27,2 50,1				
51,2 73 74,1				
75,2	0	0	0	0
52,3	0	0	-2J	2J
28,3 76,3	2	-2	0	0
4,3	0	0	2J	-2J

S

	1	2	3	4
1	2	2	2	2
49	2	2	-2	-2
2,1 4,3 25				
26,1 28,3 50,1				
52,3 73 74,1				
76,3	0	0	0	0
3,2	0	0	2	-2
27,2 75,2	2J	-2J	0	0
51,2	0	0	-2	2

D

	1	2	3	4	5
1	2	1	1	1	1
49	2	-1	-1	-1	-1
3,2	0	-1,T	-1,T	1,T	1,T
76,3	0	1	-1	1	-1
26,1	0	1,TJ	-1,TJ	-1,TJ	1,TJ
51,2	0	1,T	1,T	-1,T	-1,T
28,3	0	-1	1	-1	1
74,1	0	-1,TJ	1,TJ	1,TJ	-1,TJ

C

	1	2	3	4	5
1	1	1	1	1	2
49	1	1	1	1	-2
2,1 50,1	1	1	-1	-1	0
27,2 75,2	-1,TJ	1,TJ	-1,TJ	1,TJ	0
28,3 76,3	1,T	-1,T	-1,T	1,T	0

V

	1	2	3	4
1	1	1	1	1
49	1	1	-1	-1
28,3	1,T	-1,T	-1,TJ	1,TJ
76,3	1,T	-1,T	1,TJ	-1,TJ

GM

	1+	2+	3+	4+	1-	2-	3-	4-	5+	5-
1	1	1	1	1	1	1	1	1	2	2
49	1	1	1	1	1	1	1	1	-2	-2
4 52	1	1	-1	-1	1	1	-1	-1	0	0
2,1 50,1	1	-1	1	-1	1	-1	1	-1	0	0
3,1 51,1	1	-1	-1	1	1	-1	-1	1	0	0
26,1 74,1	1	-1	1	-1	-1	1	-1	1	0	0
27,1 75,1	1	-1	-1	1	-1	1	1	-1	0	0
73	1	1	1	1	-1	-1	-1	-1	-2	2
28 76	1	1	-1	-1	-1	-1	1	1	0	0
25	1	1	1	1	-1	-1	-1	-1	2	-2

DT

	1	2	3	4	5
1	1	1	1	1	2
49	1	1	1	1	-2
3,1 51,1	1,T	1,T	-1,T	-1,T	0
26,1 74,1	1,T	-1,T	1,T	-1,T	0
28 76	1	-1	-1	1	0

SM

	1	2	3	4	5
1	1	1	1	1	2
49	1	1	1	1	-2
2,1 50,1	1,T	1,T	-1,T	-1,T	0
27,1 75,1	1,T	-1,T	1,T	-1,T	0
28 76	1	-1	-1	1	0

X

	1	2	3	4
1	2	2	2	2
49	2	2	-2	-2
4	0	0	2J	-2J
2,1 3,1 25				
26,1 27,1 50,1				
51,1 73 74,1				
75,1	0	0	0	0
28 76	2	-2	0	0
52	0	0	-2J	2J

Y

	1	2	3	4
1	2	2	2	2
49	2	2	-2	-2
4	0	0	2J	-2J
2,1 3,1 25				
26,1 27,1 50,1				
51,1 73 74,1				
75,1	0	0	0	0
28 76	-2	2	0	0
52	0	0	-2J	2J

S	1+	2+	3+	4+	1-	2-	3-	4-	5+	5-
1	1	1	1	1	1	1	1	1	2	2
49	1	1	1	1	1	1	1	1	-2	-2
4 52	1	1	-1	-1	1	1	-1	-1	0	0
2,1 50,1	J	-J	J	-J	J	-J	J	-J	0	0
3,1 51,1	J	-J	-J	J	J	-J	-J	J	0	0
26,1 74,1	J	-J	J	-J	-J	J	-J	J	0	0
27,1 75,1	J	-J	-J	J	-J	J	J	-J	0	0
73	1	1	1	1	-1	-1	-1	-1	-2	2
28 76	1	1	-1	-1	-1	-1	1	1	0	0
25	1	1	1	1	-1	-1	-1	-1	2	-2

D	1	2	3	4	5
1	1	1	1	1	2
49	1	1	1	1	-2
3,1 51,1	-1,TJ	-1,TJ	1,TJ	1,TJ	0
26,1 74,1	-1,TJ	1,TJ	-1,TJ	1,TJ	0
28 76	1	-1	-1	1	0

C	1	2	3	4	5
1	1	1	1	1	2
49	1	1	1	1	-2
2,1 50,1	-1,TJ	-1,TJ	1,TJ	1,TJ	0
27,1 75,1	-1,TJ	1,TJ	-1,TJ	1,TJ	0
28 76	1	-1	-1	1	0

V	1	2	3	4
1	1	1	1	1
49	1	1	-1	-1
28	1	-1	J	-J
76	1	-1	-J	J

GM

GM	1+	2+	3+	4+	1-	2-	3-	4-	5+	5-
1	1	1	1	1	1	1	1	1	2	2
49	1	1	1	1	1	1	1	1	-2	-2
3,1 51,1	1	1	-1	-1	1	1	-1	-1	0	0
4,2 52,2	1	-1	1	-1	1	1	-1	-1	0	0
2,3 50,3	1	-1	-1	1	1	-1	-1	1	0	0
28,2 76,2	1	-1	1	-1	-1	1	-1	1	0	0
26,3 74,3	1	-1	-1	1	-1	1	1	-1	0	0
73	1	1	1	1	-1	-1	-1	-1	-2	2
27,1 75,1	1	1	-1	-1	-1	-1	1	1	0	0
25	1	1	1	1	-1	-1	-1	-1	2	-2

DT

DT	1	2	3	4	5
1	1	1	1	1	2
49	1	1	1	1	-2
3,1 51,1	1,T	1,T	-1,T	-1,T	0
28,2 76,2	1	-1	1	-1	0
26,3 74,3	1,T	-1,T	-1,T	1,T	0

SM

SM	1	2	3	4	5
1	1	1	1	1	2
49	1	1	1	1	-2
2,3 50,3	1,T	1,T	-1,T	-1,T	0
28,2 76,2	1,T	-1,T	1,T	-1,T	0
27,1 75,1	1	-1	-1	1	0

X

X	1	2	3	4
1	2	2	2	2
49	2	2	-2	-2
3,1	0	0	2J	-2J
2,3 4,2 25				
26,3 28,2 50,3				
52,2 73 74,3				
76,2	0	0	0	0
27,1 75,1	-2	2	0	0
51,1	0	0	-2J	2J

Y

Y	1	2	3	4
1	2	2	2	2
49	2	2	-2	-2
2,3 3,1 25				
26,3 27,1 50,3				
51,1 73 74,3				
75,1	0	0	0	0
4,2	0	0	2J	-2J
28,2 76,2	2	-2	0	0
52,2	0	0	-2J	2J

S

	1	2	3	4
1	2	2	2	2
49	2	2	-2	-2

3,1 4,2 25
27,1 28,2 51,1
52,2 73 75,1

	1	2	3	4
76,2	0	0	0	0
74,3	0	0	2	-2
2,3 50,3	-2J	2J	0	0
26,3	0	0	-2	2

D

	1	2	3	4	5
1	2	1	1	1	1
49	2	-1	-1	-1	-1
3,1	0	-1,TJ	-1,TJ	1,TJ	1,TJ
28,2	0	1	-1	1	-1
74,3	0	-1,T	1,T	1,T	-1,T
51,1	0	1,TJ	1,TJ	-1,TJ	-1,TJ
76,2	0	-1	1	-1	1
26,3	0	1,T	-1,T	-1,T	1,T

C

	1	2	3	4	5
1	1	1	1	1	2
49	1	1	1	1	-2
2,3 50,3	-1,TJ	-1,TJ	1,TJ	1,TJ	0
28,2 76,2	1,T	-1,T	1,T	-1,T	0
27,1 75,1	1	-1	-1	1	0

V

	1	2	3	4
1	1	1	1	1
49	1	1	-1	-1
28,2	1,T	-1,T	-1,TJ	1,TJ
76,2	1,T	-1,T	1,TJ	-1,TJ

GM

	1+	2+	3+	4+	1-	2-	3	4-	5+	5-
1	1	1	1	1	1	1	1	1	2	2
49	1	1	1	1	1	1	1	1	-2	-2
4 52	1	1	-1	-1	1	1	-1	-1	0	0
2,1 50,1	1	-1	1	-1	1	-1	1	-1	0	0
3,1 51,1	1	-1	-1	1	1	-1	-1	1	0	0
26 74	1	-1	1	-1	-1	1	-1	1	0	0
27 75	1	-1	-1	1	-1	1	1	-1	0	0
73,1	1	1	1	1	-1	-1	-1	-1	-2	2
28,1 76,1	1	1	-1	-1	-1	-1	1	1	0	0
25,1	1	1	1	1	-1	-1	-1	-1	2	-2

DT

	1	2	3	4	5
1	1	1	1	1	2
49	1	1	1	1	-2
3,1 51,1	1,T	1,T	-1,T	-1,T	0
26 74	1	-1	1	-1	0
28,1 76,1	1,T	-1,T	-1,T	1,T	0

SM

	1	2	3	4	5
1	1	1	1	1	2
49	1	1	1	1	-2
2,1 50,1	1,T	1,T	-1,T	-1,T	0
27 75	1	-1	1	-1	0
28,1 76,1	1,T	-1,T	-1,T	1,T	0

X

	1	2	3	4
1	2	2	2	2
49	2	2	-2	-2
2,1 4 25,1 26 28,1 50,1 52 73,1 74 76,1	0	0	0	0
51,1	0	0	2J	-2J
27 75	2	-2	0	0
3,1	0	0	-2J	2J

Y

	1	2	3	4
1	2	2	2	2
49	2	2	-2	-2
3,1 4 25,1 27 28,1 51,1 52 73,1 75 76,1	0	0	0	0
2,1	0	0	2J	-2J
26 74	2	-2	0	0
50,1	0	0	-2J	2J

S

	1	2	3	4
1	2	2	2	2
49	2	2	-2	-2
4 52	2	-2	0	0
2,1 3,1 25,1				
26 27 50,1				
51,1 73,1 74				
75	0	0	0	0
76,1	0	0	-2J	2J
28,1	0	0	2J	-2J

D

	1	2	3	4	5
1	2	1	1	1	1
49	2	-1	-1	-1	-1
51,1	0	-1,T	-1,T	1,T	1,T
26	0	J	-J	J	-J
76,1	0	1,TJ	-1,TJ	-1,TJ	1,TJ
3,1	0	1,T	1,T	-1,T	-1,T
74	0	-J	J	-J	J
28,1	0	-1,TJ	1,TJ	1,TJ	-1,TJ

C

	1	2	3	4	5
1	2	1	1	1	1
49	2	-1	-1	-1	-1
2,1	0	-1,T	-1,T	1,T	1,T
75	0	J	-J	J	-J
28,1	0	1,TJ	-1,TJ	-1,TJ	1,TJ
50,1	0	1,T	1,T	-1,T	-1,T
27	0	-J	J	-J	J
76,1	0	-1,TJ	1,TJ	1,TJ	-1,TJ

V

	1	2	3	4
1	1	1	1	1
49	1	1	-1	-1
28,1	1,T	-1,T	-1,TJ	1,TJ
76,1	1,T	-1,T	1,TJ	-1,TJ

GM

	1+	2+	3+	4+	1-	2-	3-	4-	5+	5-
1	1	1	1	1	1	1	1	1	2	2
49	1	1	1	1	1	1	1	1	-2	-2
4 52	1	1	-1	-1	1	1	-1	-1	0	0
2 50	1	-1	1	-1	1	-1	1	-1	0	0
3 51	1	-1	-1	1	1	-1	-1	1	0	0
26 74	1	-1	1	-1	-1	1	-1	1	0	0
27 75	1	-1	-1	1	-1	1	1	-1	0	0
73	1	1	1	1	-1	-1	-1	-1	-2	2
28 76	1	1	-1	-1	-1	-1	1	1	0	0
25	1	1	1	1	-1	-1	-1	-1	2	-2

DT

	1	2	3	4	5
1	1	1	1	1	2
49	1	1	1	1	-2
3 51	1	1	-1	-1	0
26 74	1	-1	1	-1	0
28 76	1	-1	-1	1	0

SM

	1	2	3	4	5
1	1	1	1	1	2
49	1	1	1	1	-2
2 50	1	1	-1	-1	0
27 75	1	-1	1	-1	0
28 76	1	-1	-1	1	0

YA

	1+	2+	3+	4+	1-	2-	3-	4-	5+	5-
1	1	1	1	1	1	1	1	1	2	2
49	1	1	1	1	1	1	1	1	-2	-2
4 52	1	1	-1	-1	1	1	-1	-1	0	0
2 50	1	-1	1	-1	1	-1	1	-1	0	0
3 51	1	-1	-1	1	1	-1	-1	1	0	0
26 74	1	-1	1	-1	-1	1	-1	1	0	0
27 75	1	-1	-1	1	-1	1	1	-1	0	0
73	1	1	1	1	-1	-1	-1	-1	-2	2
28 76	1	1	-1	-1	-1	-1	1	1	0	0
25	1	1	1	1	-1	-1	-1	-1	2	-2

YB

	1+	2+	3+	4+	1-	2-	3-	4-	5+	5-
1	1	1	1	1	1	1	1	1	2	2
49	1	1	1	1	1	1	1	1	-2	-2
4 52	1	1	-1	-1	1	1	-1	-1	0	0
2 50	1	-1	1	-1	1	-1	1	-1	0	0
3 51	1	-1	-1	1	1	-1	-1	1	0	0
26 74	1	-1	1	-1	-1	1	-1	1	0	0
27 75	1	-1	-1	1	-1	1	1	-1	0	0
73	1	1	1	1	-1	-1	-1	-1	-2	2
28 76	1	1	-1	-1	-1	-1	1	1	0	0
25	1	1	1	1	-1	-1	-1	-1	2	-2

S

	1+	2+	1−	2−	3+	4+	3−	4−
1	1	1	1	1	1	1	1	1
49	1	1	1	1	−1	−1	−1	−1
4	1	−1	1	−1	J	−J	J	−J
76	1	−1	−1	1	−J	J	J	−J
25	1	1	−1	−1	1	1	−1	−1
52	1	−1	1	−1	−J	J	−J	J
28	1	−1	−1	1	J	−J	−J	J
73	1	1	−1	−1	−1	−1	1	1

C

	1	2	3	4	5
1	1	1	1	1	2
49	1	1	1	1	−2
2 50	1	1	−1	−1	0
27 75	1	−1	1	−1	0
28 76	1	−1	−1	1	0

F

	1	2	3	4	5
1	1	1	1	1	2
49	1	1	1	1	−2
3 51	1	1	−1	−1	0
26 74	1	−1	1	−1	0
28 76	1	−1	−1	1	0

P

	1	2	3	4
1	1	1	1	1
49	1	1	−1	−1
28	1	−1	J	−J
76	1	−1	−J	J

GM

	1+	2+	3+	4+	1-	2-	3-	4-	5+	5-
1	1	1	1	1	1	1	1	1	2	2
49	1	1	1	1	1	1	1	1	-2	-2
4,1 52,1	1	1	-1	-1	1	1	-1	-1	0	0
2 50	1	-1	1	-1	1	-1	1	-1	0	0
3,1 51,1	1	-1	-1	1	1	-1	-1	1	0	0
26 74	1	-1	1	-1	-1	1	-1	1	0	0
27,1 75,1	1	-1	-1	1	-1	1	1	-1	0	0
73	1	1	1	1	-1	-1	-1	-1	-2	2
28,1 76,1	1	1	-1	-1	-1	-1	1	1	0	0
25	1	1	1	1	-1	-1	-1	-1	2	-2

DT

	1	2	3	4	5
1	1	1	1	1	2
49	1	1	1	1	-2
3,1 51,1	1	1	-1	-1	0
26 74	1	-1	1	-1	0
28,1 76,1	1	-1	-1	1	0

SM

	1	2	3	4	5
1	1	1	1	1	2
49	1	1	1	1	-2
2 50	1	1	-1	-1	0
27,1 75,1	1,T	-1,T	1,T	-1,T	0
28,1 76,1	1,T	-1,T	-1,T	1,T	0

YA

	1+	2+	3+	4+	1-	2-	3-	4-	5+	5-
1	1	1	1	1	1	1	1	1	2	2
49	1	1	1	1	1	1	1	1	-2	-2
4,1 52,1	1	1	-1	-1	1	1	-1	-1	0	0
2 50	1	-1	1	-1	1	-1	1	-1	0	0
3,1 51,1	1	-1	-1	1	1	-1	-1	1	0	0
26 74	1	-1	1	-1	-1	1	-1	1	0	0
27,1 75,1	1	-1	-1	1	-1	1	1	-1	0	0
73	1	1	1	1	-1	-1	-1	-1	-2	2
28,1 76,1	1	1	-1	-1	-1	-1	1	1	0	0
25	1	1	1	1	-1	-1	-1	-1	2	-2

YB

	1+	2+	3+	4+	1-	2-	3-	4-	5+	5-
1	1	1	1	1	1	1	1	1	2	2
49	1	1	1	1	1	1	1	1	-2	-2
4,1 52,1	1	1	-1	-1	1	1	-1	-1	0	0
2 50	1	-1	1	-1	1	-1	1	-1	0	0
3,1 51,1	1	-1	-1	1	1	-1	-1	1	0	0
26 74	1	-1	1	-1	-1	1	-1	1	0	0
27,1 75,1	1	-1	-1	1	-1	1	1	-1	0	0
73	1	1	1	1	-1	-1	-1	-1	-2	2
28,1 76,1	1	1	-1	-1	-1	-1	1	1	0	0
25	1	1	1	1	-1	-1	-1	-1	2	-2

S

	1	2
1	2	2
49	2	-2
4,1 25 28,1		
52,1 73 76,1	0	0

C

	1	2	3	4	5
1	1	1	1	1	2
49	1	1	1	1	-2
2 50	1	1	-1	-1	0
27,1 75,1	1,T	-1,T	1,T	-1,T	0
28,1 76,1	1,T	-1,T	-1,T	1,T	0

F

	1	2	3	4	5
1	1	1	1	1	2
49	1	1	1	1	-2
3,1 51,1	1	1	-1	-1	0
26 74	1	-1	1	-1	0
28,1 76,1	1	-1	-1	1	0

P

	1	2	3	4
1	1	1	1	1
49	1	1	-1	-1
28,1	1,T	-1,T	-1,TJ	1,TJ
76,1	1,T	-1,T	1,TJ	-1,TJ

GM

	1	2	3	4	5	6	7	8
1	1	1	1	1	1	1	1	1
49	1	1	1	1	-1	-1	-1	-1
4	1	1	-1	-1	J	J	-J	-J
14	1	-1	J	-J	V	-V	V*	-V*
15	1	-1	-J	J	-V*	V*	-V	V
52	1	1	-1	-1	-J	-J	J	J
62	1	-1	J	-J	-V	V	-V*	V*
63	1	-1	-J	J	V*	-V*	V	-V

M

	1	2	3	4	5	6	7	8
1	1	1	1	1	1	1	1	1
49	1	1	1	1	-1	-1	-1	-1
4	1	1	-1	-1	J	J	-J	-J
14	1	-1	J	-J	V	-V	V*	-V*
15	1	-1	-J	J	-V*	V*	-V	V
52	1	1	-1	-1	-J	-J	J	J
62	1	-1	J	-J	-V	V	-V*	V*
63	1	-1	-J	J	V*	-V*	V	-V

X

	1	2	3	4
1	1	1	1	1
49	1	1	-1	-1
4	1	-1	J	-J
52	1	-1	-J	J

GM

	1	2	3	4	5	6	7	8
1	1	1	1	1	1	1	1	1
49	1	1	1	1	-1	-1	-1	-1
4	1	1	-1	-1	J	J	-J	-J
38	1	-1	J	-J	V	-V	V*	-V*
39	1	-1	-J	J	-V*	V*	-V	V
52	1	1	-1	-1	-J	-J	J	J
86	1	-1	J	-J	-V	V	-V*	V*
87	1	-1	-J	J	V*	-V*	V	-V

M

	1	2	3	4	5	6	7	8
1	1	1	1	1	1	1	1	1
49	1	1	1	1	-1	-1	-1	-1
4	1	1	-1	-1	J	J	-J	-J
38	1	-1	J	-J	V	-V	V*	-V*
39	1	-1	-J	J	-V*	V*	-V	V
52	1	1	-1	-1	-J	-J	J	J
86	1	-1	J	-J	-V	V	-V*	V*
87	1	-1	-J	J	V*	-V*	V	-V

X

	1	2	3	4
1	1	1	1	1
49	1	1	-1	-1
4	1	-1	J	-J
52	1	-1	-J	J

P4/M

GM

	1+	2+	3+	4+	1-	2-	3-	4-	5+	6+	7+	8+	5-	6-	7-	8-
1	1	1	1	1	1	1	1	1	1	1	1	1	1	1	1	1
49	1	1	1	1	1	1	1	1	-1	-1	-1	-1	-1	-1	-1	-1
4	1	1	-1	-1	1	1	-1	-1	J	J	-J	-J	J	J	-J	-J
28	1	1	-1	-1	-1	-1	1	1	J	J	-J	-J	-J	-J	J	J
73	1	1	1	1	-1	-1	-1	-1	-1	-1	-1	-1	1	1	1	1
14	1	-1	J	-J	1	-1	J	-J	V	-V	V*	-V*	V	-V	V*	-V*
15	1	-1	-J	J	1	-1	-J	J	-V*	V*	-V	V	-V*	V*	-V	V
39	1	-1	-J	J	-1	1	J	-J	-V*	V*	-V	V	V*	-V*	V	-V
86	1	-1	J	-J	-1	1	-J	J	-V	V	-V*	V*	V	-V	V*	-V*
52	1	1	-1	-1	1	1	-1	-1	-J	-J	J	J	-J	-J	J	J
76	1	1	-1	-1	-1	-1	1	1	-J	-J	J	J	J	J	-J	-J
25	1	1	1	1	-1	-1	-1	-1	1	1	1	1	-1	-1	-1	-1
62	1	-1	J	-J	1	-1	J	-J	-V	V	-V*	V*	-V	V	-V*	V*
63	1	-1	-J	J	1	-1	-J	J	V*	-V*	V	-V	V*	-V*	V	-V
87	1	-1	-J	J	-1	1	J	-J	V*	-V*	V	-V	-V*	V*	-V	V
38	1	-1	J	-J	-1	1	-J	J	V	-V	V*	-V*	-V	V	-V*	V*

SM

	1	2	3	4
1	1	1	1	1
49	1	1	-1	-1
28	1	-1	J	-J
76	1	-1	-J	J

DT

	1	2	3	4
1	1	1	1	1
49	1	1	-1	-1
28	1	-1	J	-J
76	1	-1	-J	J

M

	1+	2+	3+	4+	1-	2-	3-	4-	5+	6+	7+	8+	5-	6-	7-	8-
1	1	1	1	1	1	1	1	1	1	1	1	1	1	1	1	1
49	1	1	1	1	1	1	1	1	-1	-1	-1	-1	-1	-1	-1	-1
4	1	1	-1	-1	1	1	-1	-1	J	J	-J	-J	J	J	-J	-J
28	1	1	-1	-1	-1	-1	1	1	J	J	-J	-J	-J	-J	J	J
73	1	1	1	1	-1	-1	-1	-1	-1	-1	-1	-1	1	1	1	1
14	1	-1	J	-J	1	-1	J	-J	V	-V	V*	-V*	V	-V	V*	-V*
15	1	-1	-J	J	1	-1	-J	J	-V*	V*	-V	V	-V*	V*	-V	V
39	1	-1	-J	J	-1	1	J	-J	-V*	V*	-V	V	V*	-V*	V	-V
86	1	-1	J	-J	-1	1	-J	J	-V	V	-V*	V*	V	-V	V*	-V*
52	1	1	-1	-1	1	1	-1	-1	-J	-J	J	J	-J	-J	J	J
76	1	1	-1	-1	-1	-1	1	1	-J	-J	J	J	J	J	-J	-J
25	1	1	1	1	-1	-1	-1	-1	1	1	1	1	-1	-1	-1	-1
62	1	-1	J	-J	1	-1	J	-J	-V	V	-V*	V*	-V	V	-V*	V*
63	1	-1	-J	J	1	-1	-J	J	V*	-V*	V	-V	V*	-V*	V	-V
87	1	-1	-J	J	-1	1	J	-J	V*	-V*	V	-V	-V*	V*	-V	V
38	1	-1	J	-J	-1	1	-J	J	V	-V	V*	-V*	-V	V	-V*	V*

X	1+	2+	1-	2-	3+	4+	3-	4-
1	1	1	1	1	1	1	1	1
49	1	1	1	1	-1	-1	-1	-1
4	1	-1	1	-1	J	-J	J	-J
28	1	-1	-1	1	J	-J	-J	J
73	1	1	-1	-1	-1	-1	1	1
52	1	-1	1	-1	-J	J	-J	J
76	1	-1	-1	1	-J	J	J	-J
25	1	1	-1	-1	1	1	-1	-1

Y	1	2	3	4
1	1	1	1	1
49	1	1	-1	-1
28	1	-1	J	-J
76	1	-1	-J	J

D	1	2	3	4
1	1	1	1	1
49	1	1	-1	-1
28	1	-1	J	-J
76	1	-1	-J	J

DA	1	2	3	4
1	1	1	1	1
49	1	1	-1	-1
28	1	-1	-J	J
76	1	-1	J	-J

GM

	1+	2+	3+	4+	1-	2-	3-	4-	5+	6+	7+	8+	5-	6-	7-	8-
1	1	1	1	1	1	1	1	1	1	1	1	1	1	1	1	1
49	1	1	1	1	1	1	1	1	-1	-1	-1	-1	-1	-1	-1	-1
4	1	1	-1	-1	1	1	-1	-1	J	J	-J	-J	J	J	-J	-J
28,1	1	1	-1	-1	-1	-1	1	1	J	J	-J	-J	-J	-J	J	J
73,1	1	1	1	1	-1	-1	-1	-1	-1	-1	-1	-1	1	1	1	1
14,1	1	-1	J	-J	1	-1	J	-J	V	-V	V*	-V*	V	-V	V*	-V*
15,1	1	-1	-J	J	1	-1	-J	J	-V*	V*	-V	V	-V*	V*	-V	V
39	1	-1	-J	J	-1	1	J	-J	-V*	V*	-V	V	V*	-V*	-V	V
86	1	-1	J	-J	-1	1	-J	J	-V	V	-V*	V*	V	-V	V*	-V*
52	1	1	-1	-1	1	1	-1	-1	-J	-J	J	J	-J	-J	J	J
76,1	1	1	-1	-1	-1	-1	1	1	-J	-J	J	J	J	J	-J	-J
25,1	1	1	1	1	-1	-1	-1	-1	1	1	1	1	-1	-1	-1	-1
62,1	1	-1	J	-J	1	-1	J	-J	-V	V	-V*	V*	-V	V	-V*	V*
63,1	1	-1	-J	J	1	-1	-J	J	V*	-V*	V	-V	V*	-V*	V	-V
87	1	-1	-J	J	-1	1	J	-J	V*	-V*	V	-V	-V*	V*	-V	V
38	1	-1	J	-J	-1	1	-J	J	V	-V	V*	-V*	-V	V	-V*	V*

SM

	1	2	3	4
1	1	1	1	1
49	1	1	-1	-1
28,1	1,T	-1,T	-1,TJ	1,TJ
76,1	1,T	-1,T	1,TJ	-1,TJ

DT

	1	2	3	4
1	1	1	1	1
49	1	1	-1	-1
28,1	1,T	-1,T	-1,TJ	1,TJ
76,1	1,T	-1,T	1,TJ	-1,TJ

M

	1	2	3	4
1	2	2	2	2
49	2	2	-2	-2
4	2	-2	2J	-2J
14,1 15,1 25,1				
28,1 38 39				
62,1 63,1 73,1				
76,1 86 87	0	0	0	0
52	2	-2	-2J	2J

X

	1	2
1	2	2
49	2	-2
4 25,1 28,1		
52 73,1 76,1	0	0

D

	1	2	3	4
1	1	1	1	1
49	1	1	-1	-1
28,1	1,T	-1,T	-1,TJ	1,TJ
76,1	1,T	-1,T	1,TJ	-1,TJ

Y

	1	2	3	4
1	1	1	1	1
49	1	1	-1	-1
28,1	-1,T	1,T	-1,TJ	1,TJ
76,1	-1,T	1,T	1,TJ	-1,TJ

DA

	1	2	3	4
1	1	1	1	1
49	1	1	-1	-1
28,1	1,T	-1,T	1,TJ	-1,TJ
76,1	1,T	-1,T	-1,TJ	1,TJ

GM

	1	2	3	4	5	6	7
1	1	1	1	1	2	2	2
49	1	1	1	1	2	-2	-2
4 52	1	1	1	1	-2	0	0
2 3 50 51	1	1	-1	-1	0	0	0
14 63	1	-1	1	-1	0	A	-A
13 16 61 64	1	-1	-1	-1	0	0	0
62 15	1	-1	1	-1	0	-A	A

SM

	1	2	3	4
1	1	1	1	1
49	1	1	-1	-1
16	1	-1	J	-J
64	1	-1	-J	J

DT

	1	2	3	4
1	1	1	1	1
49	1	1	-1	-1
51	1	-1	J	-J
3	1	-1	-J	J

M

	1	2	3	4	5	6	7
1	1	1	1	1	2	2	2
49	1	1	1	1	2	-2	-2
4 52	1	1	1	1	-2	0	0
2 3 50 51	1	1	-1	-1	0	0	0
14 63	1	-1	1	-1	0	A	-A
13 16 61 64	1	-1	-1	1	0	0	0
62 15	1	-1	1	-1	0	-A	A

X

	1	2	3	4	5
1	1	1	1	1	2
49	1	1	1	1	-2
4 52	1	1	-1	-1	0
2 50	1	-1	1	-1	0
3 51	1	-1	-1	1	0

Y

	1	2	3	4
1	1	1	1	1
49	1	1	-1	-1
2	1	-1	J	-J
50	1	-1	-J	J

GM

	1	2	3	4	5	6	7
1	1	1	1	1	2	2	2
49	1	1	1	1	2	-2	-2
4 52	1	1	1	1	-2	0	0
2,1 3,1 50,1							
51,1	1	1	-1	-1	0	0	0
14,1 63,1	1	-1	1	-1	0	A	-A
13 16 61 64	1	-1	-1	1	0	0	0
62,1 15,1	1	-1	1	-1	0	-A	A

SM

	1	2	3	4
1	1	1	1	1
49	1	1	-1	-1
16	1	-1	J	-J
64	1	-1	-J	J

DT

	1	2	3	4
1	1	1	1	1
49	1	1	-1	-1
51,1	1,T	-1,T	-1,TJ	1,TJ
3,1	1,T	-1,T	1,TJ	-1,TJ

M

	1	2	3	4	5	6	7
1	1	1	1	1	2	2	2
49	1	1	1	1	2	-2	-2
4 52	1	1	1	1	-2	0	0
2,1 3,1 50,1							
51,1	J	J	-J	-J	0	0	0
14,1 63,1	J	-J	J	-J	0	-AJ	AJ
13 16 61 64	1	-1	-1	1	0	0	0
62,1 15,1	J	-J	J	-J	0	AJ	-AJ

X

	1	2	3	4	5
1	2	1	1	1	1
49	2	-1	-1	-1	-1
4	0	J	J	-J	-J
2,1	0	J	-J	J	-J
51,1	0	-1	1	1	-1
52	0	-J	-J	J	J
50,1	0	-J	J	-J	J
3,1	0	1	-1	-1	1

Y

	1	2	3	4
1	1	1	1	1
49	1	1	-1	-1
2,1	-1,TJ	1,TJ	-1,T	1,T
50,1	-1,TJ	1,TJ	1,T	-1,T

GM

	1	2	3	4	5	6	7
1	1	1	1	1	2	2	2
49	1	1	1	1	2	-2	-2
4 52	1	1	1	1	-2	0	0
26 27 74 75	1	1	-1	-1	0	0	0
37 40 85 88	1	-1	1	-1	0	0	0
15 62	1	-1	-1	1	0	A	-A
63 14	1	-1	-1	1	0	-A	A

SM

	1	2	3	4
1	1	1	1	1
49	1	1	-1	-1
85	1	-1	J	-J
37	1	-1	-J	J

DT

	1	2	3	4
1	1	1	1	1
49	1	1	-1	-1
26	1	-1	J	-J
74	1	-1	-J	J

M

	1	2	3	4	5	6	7
1	1	1	1	1	2	2	2
49	1	1	1	1	2	-2	-2
4 52	1	1	1	1	-2	0	0
26 27 74 75	1	1	-1	-1	0	0	0
37 40 85 88	1	-1	1	-1	0	0	0
15 62	1	-1	-1	1	0	A	-A
63 14	1	-1	-1	1	0	-A	A

X

	1	2	3	4	5
1	1	1	1	1	2
49	1	1	1	1	-2
4 52	1	1	-1	-1	0
26 74	1	-1	1	-1	0
27 75	1	-1	-1	1	0

Y

	1	2	3	4
1	1	1	1	1
49	1	1	-1	-1
75	1	-1	J	-J
27	1	-1	-J	J

GM

	1	2	3	4	5	6	7
1	1	1	1	1	2	2	2
49	1	1	1	1	2	-2	-2
4 52	1	1	1	1	-2	0	0
26,1 27,1 74,1							
75,1	1	1	-1	-1	0	0	0
37,1 40,1 85,1							
88,1	1	-1	1	-1	0	0	0
15 62	1	-1	-1	1	0	A	-A
63 14	1	-1	-1	1	0	-A	A

SM

	1	2	3	4
1	1	1	1	1
49	1	1	-1	-1
85,1	1,T	-1,T	-1,TJ	1,TJ
37,1	1,T	-1,T	1,TJ	-1,TJ

DT

	1	2	3	4
1	1	1	1	1
49	1	1	-1	-1
26,1	1,T	-1,T	-1,TJ	1,TJ
74,1	1,T	-1,T	1,TJ	-1,TJ

M

	1	2	3	4	5	6	7
1	1	1	1	1	2	2	2
49	1	1	1	1	2	-2	-2
4 52	-1	-1	-1	-1	2	0	0
26,1 74,1	J	J	-J	-J	0	0	0
27,1 75,1	-J	-J	J	J	0	0	0
40,1 88,1	1	-1	1	-1	0	0	0
37,1 85,1	-1	1	-1	1	0	0	0
62	J	-J	-J	J	0	AJ	-AJ
63	-J	J	J	-J	0	AJ	-AJ
14	J	-J	-J	J	0	-AJ	AJ
15	-J	J	J	-J	0	-AJ	AJ

X

	1	2	3	4	5
1	2	1	1	1	1
49	2	-1	-1	-1	-1
4	0	J	J	-J	-J
26,1	0	1	-1	1	-1
75,1	0	J	-J	-J	J
52	0	-J	-J	J	J
74,1	0	-1	1	-1	1
27,1	0	-J	J	J	-J

Y

	1	2	3	4
1	1	1	1	1
49	1	1	-1	-1
75,1	-1,TJ	1,TJ	-1,T	1,T
27,1	-1,TJ	1,TJ	1,T	-1,T

GM

	1	2	3	4	5	6	7
1	1	1	1	1	2	2	2
49	1	1	1	1	2	-2	-2
4 52	1	1	1	1	-2	0	0
2 3 50 51	1	1	-1	-1	0	0	0
37 40 85 88	1	-1	1	-1	0	0	0
39 86	1	-1	-1	1	0	A	-A
87 38	1	-1	-1	1	0	-A	A

SM

	1	2	3	4
1	1	1	1	1
49	1	1	-1	-1
85	1	-1	J	-J
37	1	-1	-J	J

DT

	1	2	3	4
1	1	1	1	1
49	1	1	-1	-1
51	1	-1	J	-J
3	1	-1	-J	J

M

	1	2	3	4	5	6	7
1	1	1	1	1	2	2	2
49	1	1	1	1	2	-2	-2
4 52	1	1	1	1	-2	0	0
2 3 50 51	1	1	-1	-1	0	0	0
37 40 85 88	1	-1	1	-1	0	0	0
39 86	1	-1	-1	1	0	A	-A
87 38	1	-1	-1	1	0	-A	A

X

	1	2	3	4	5
1	1	1	1	1	2
49	1	1	1	1	-2
4 52	1	1	-1	-1	0
2 50	1	-1	1	-1	0
3 51	1	-1	-1	1	0

Y

	1	2	3	4
1	1	1	1	1
49	1	1	-1	-1
2	1	-1	J	-J
50	1	-1	-J	J

GM

	1	2	3	4	5	6	7
1	1	1	1	1	2	2	2
49	1	1	1	1	2	-2	-2
4 52	1	1	1	1	-2	0	0
2,1 3,1 50,1							
51,1	1	1	-1	-1	0	0	0
37,1 40,1 85,1							
88,1	1	-1	1	-1	0	0	0
39 86	1	-1	-1	1	0	A	-A
87 38	1	-1	-1	1	0	-A	A

SM

	1	2	3	4
1	1	1	1	1
49	1	1	-1	-1
85,1	1,T	-1,T	-1,TJ	1,TJ
37,1	1,T	-1,T	1,TJ	-1,TJ

DT

	1	2	3	4
1	1	1	1	1
49	1	1	-1	-1
51,1	1,T	-1,T	-1,TJ	1,TJ
3,1	1,T	-1,T	1,TJ	-1,TJ

M

	1	2	3	4	5	6	7
1	1	1	1	1	2	2	2
49	1	1	1	1	2	-2	-2
4 52	-1	-1	-1	-1	2	0	0
2,1 50,1	J	J	-J	-J	0	0	0
3,1 51,1	-J	-J	J	J	0	0	0
40,1 88,1	1	-1	1	-1	0	0	0
37,1 85,1	-1	1	-1	1	0	0	0
86	J	-J	-J	J	0	AJ	-AJ
87	-J	J	J	-J	0	AJ	-AJ
38	J	-J	-J	J	0	-AJ	AJ
39	-J	J	J	-J	0	-AJ	AJ

X

	1	2	3	4	5
1	2	1	1	1	1
49	2	-1	-1	-1	-1
4	0	J	J	-J	-J
2,1	0	J	-J	J	-J
51,1	0	-1	1	1	-1
52	0	-J	-J	J	J
50,1	0	-J	J	-J	J
3,1	0	1	-1	-1	1

Y

	1	2	3	4
1	1	1	1	1
49	1	1	-1	-1
2,1	-1,TJ	1,TJ	-1,T	1,T
50,1	-1,TJ	1,TJ	1,T	-1,T

GM

	1	2	3	4	5	6	7
1	1	1	1	1	2	2	2
49	1	1	1	1	2	-2	-2
4 52	1	1	1	1	-2	0	0
26 27 74 75	1	1	-1	-1	0	0	0
38 87	1	-1	1	-1	0	A	-A
13 16 61 64	1	-1	-1	1	0	0	0
86 39	1	-1	1	-1	0	-A	A

SM

	1	2	3	4
1	1	1	1	1
49	1	1	-1	-1
16	1	-1	J	-J
64	1	-1	-J	J

DT

	1	2	3	4
1	1	1	1	1
49	1	1	-1	-1
26	1	-1	J	-J
74	1	-1	-J	J

M

	1	2	3	4	5	6	7
1	1	1	1	1	2	2	2
49	1	1	1	1	2	-2	-2
4 52	1	1	1	1	-2	0	0
26 27 74 75	1	1	-1	-1	0	0	0
38 87	1	-1	1	-1	0	A	-A
13 16 61 64	1	-1	-1	1	0	0	0
86 39	1	-1	1	-1	0	-A	A

X

	1	2	3	4	5
1	1	1	1	1	2
49	1	1	1	1	-2
4 52	1	1	-1	-1	0
26 74	1	-1	1	-1	0
27 75	1	-1	-1	1	0

Y

	1	2	3	4
1	1	1	1	1
49	1	1	-1	-1
75	1	-1	J	-J
27	1	-1	-J	J

115

GM

	1	2	3	4	5	6	7
1	1	1	1	1	2	2	2
49	1	1	1	1	2	-2	-2
4 52	1	1	1	1	-2	0	0
26,1 27,1 74,1							
75,1	1	1	-1	-1	0	0	0
38 87	1	-1	1	-1	0	A	-A
13,1 16,1 61,1							
64,1	1	-1	-1	1	0	0	0
86 39	1	-1	1	-1	0	-A	A

SM

	1	2	3	4
1	1	1	1	1
49	1	1	-1	-1
16,1	1,T	-1,T	-1,TJ	1,TJ
64,1	1,T	-1,T	1,TJ	-1,TJ

DT

	1	2	3	4
1	1	1	1	1
49	1	1	-1	-1
26,1	1,T	-1,T	-1,TJ	1,TJ
74,1	1,T	-1,T	1,TJ	-1,TJ

M

	1	2	3	4	5	6	7
1	1	1	1	1	2	2	2
49	1	1	1	1	2	-2	-2
4 52	-1	-1	-1	-1	2	0	0
26,1 74,1	J	J	-J	-J	0	0	0
27,1 75,1	-J	-J	J	J	0	0	0
38	J	-J	J	-J	0	AJ	-AJ
39	-J	J	-J	J	0	AJ	-AJ
16,1 64,1	1	-1	-1	1	0	0	0
13,1 61,1	-1	1	1	-1	0	0	0
86	J	-J	J	-J	0	-AJ	AJ
87	-J	J	-J	J	0	-AJ	AJ

X	1	2	3	4	5
1	2	1	1	1	1
49	2	-1	-1	-1	-1
4	0	J	J	-J	-J
26,1	0	1	-1	1	-1
75,1	0	J	-J	-J	J
52	0	-J	-J	J	J
74,1	0	-1	1	-1	1
27,1	0	-J	J	J	-J

Y	1	2	3	4
1	1	1	1	1
49	1	1	-1	-1
75,1	-1,TJ	1,TJ	-1,T	1,T
27,1	-1,TJ	1,TJ	1,T	-1,T

GM	1+	2+	3+	4+	5+	1-	2-	3-	4-	5-	6+	7+	6-	7-
1	1	1	1	1	2	1	1	1	1	2	2	2	2	2
49	1	1	1	1	2	1	1	1	1	2	-2	-2	-2	-2
4														
52	1	1	1	1	-2	1	1	1	1	-2	0	0	0	0
2														
3														
50														
51	1	1	-1	-1	0	1	1	-1	-1	0	0	0	0	0
26														
27														
74														
75	1	1	-1	-1	0	-1	-1	1	1	0	0	0	0	0
73	1	1	1	1	2	-1	-1	-1	-1	-2	-2	-2	2	2
28														
76	1	1	1	1	-2	-1	-1	-1	-1	2	0	0	0	0
37														
40														
85														
88	1	-1	-1	1	0	-1	1	1	-1	0	0	0	0	0
39														
86	1	-1	1	-1	0	-1	1	-1	1	0	-A	A	A	-A
15														
62	1	-1	1	-1	0	1	-1	1	-1	0	-A	A	-A	A
13														
16														
61														
64	1	-1	-1	1	0	1	-1	-1	1	0	0	0	0	0
25	1	1	1	1	2	-1	-1	-1	-1	-2	2	2	-2	-2
87														
38	1	-1	1	-1	0	-1	1	-1	1	0	A	-A	-A	A
63														
14	1	-1	1	-1	0	1	-1	1	-1	0	A	-A	A	-A

SM		1	2	3	4	5
1		1	1	1	1	2
49		1	1	1	1	-2
28	76	1	1	-1	-1	0
37	85	1	-1	1	-1	0
16	64	1	-1	-1	1	0

DT		1	2	3	4	5
1		1	1	1	1	2
49		1	1	1	1	-2
3	51	1	1	-1	-1	0
26	74	1	-1	1	-1	0
28	76	1	-1	-1	1	0

M	1+	2+	3+	4+	5+	1-	2-	3-	4-	5-	6+	7+	6-	7-
1	1	1	1	1	2	1	1	1	1	2	2	2	2	2
49	1	1	1	1	2	1	1	1	1	2	-2	-2	-2	-2
4														
52	1	1	1	1	-2	1	1	1	1	-2	0	0	0	0
2														
3														
50														
51	1	1	-1	-1	0	1	1	-1	-1	0	0	0	0	0
26														
27														
74														
75	1	1	-1	-1	0	-1	-1	1	1	0	0	0	0	0
73	1	1	1	1	2	-1	-1	-1	-1	-2	-2	-2	2	2
28														
76	1	1	1	1	-2	-1	-1	-1	-1	2	0	0	0	0
37														
40														
85														
88	1	-1	-1	1	0	-1	1	1	-1	0	0	0	0	0
39														
86	1	-1	1	-1	0	-1	1	-1	1	0	-A	A	A	-A
15														
62	1	-1	1	-1	0	1	-1	1	-1	0	-A	A	-A	A
13														
16														
61														
64	1	-1	-1	1	0	1	-1	-1	1	0	0	0	0	0
25	1	1	1	1	2	-1	-1	-1	-1	-2	2	2	-2	-2
87														
38	1	-1	1	-1	0	-1	1	-1	1	0	A	-A	-A	A
63														
14	1	-1	1	-1	0	1	-1	1	-1	0	A	-A	A	-A

X		1+	2+	3+	4+	1-	2-	3-	4-	5+	5-
1		1	1	1	1	1	1	1	1	2	2
49		1	1	1	1	1	1	1	1	-2	-2
4	52	1	1	-1	-1	1	1	-1	-1	0	0
2	50	1	-1	1	-1	1	-1	1	-1	0	0
3	51	1	-1	-1	1	1	-1	-1	1	0	0
26	74	1	-1	1	-1	-1	1	-1	1	0	0
27	75	1	-1	-1	1	-1	1	1	-1	0	0
73		1	1	1	1	-1	-1	-1	-1	-2	2
28	76	1	1	-1	-1	-1	-1	1	1	0	0
25		1	1	1	1	-1	-1	-1	-1	2	-2

Y		1	2	3	4	5
1		1	1	1	1	2
49		1	1	1	1	-2
2	50	1	1	-1	-1	0
27	75	1	-1	1	-1	0
28	76	1	-1	-1	1	0

D		1	2	3	4
1		1	1	1	1
49		1	1	-1	-1
28		1	-1	J	-J
76		1	-1	-J	J

P4/N2/B2/M

GM	1+	2+	3+	4+	5+	1-	2-	3-	4-	5-	6+	7+	6-	7-
1	1	1	1	1	2	1	1	1	1	2	2	2	2	2
49	1	1	1	1	2	1	1	1	1	2	-2	-2	-2	-2
4														
52	1	1	1	1	-2	1	1	1	1	-2	0	0	0	0
2														
3														
50														
51	1	1	-1	-1	0	1	1	-1	-1	0	0	0	0	0
26,1														
27,1														
74,1														
75,1	1	1	-1	-1	0	-1	-1	1	1	0	0	0	0	0
73,1	1	1	1	1	2	-1	-1	-1	-1	-2	-2	-2	2	2
28,1														
76,1	1	1	1	1	-2	-1	-1	-1	-1	2	0	0	0	0
37,1														
40,1														
85,1														
88,1	1	-1	-1	1	0	-1	1	1	-1	0	0	0	0	0
39,1														
86,1	1	-1	1	-1	0	-1	1	-1	1	0	-A	A	A	-A
15														
62	1	-1	1	-1	0	1	-1	1	-1	0	-A	A	-A	A
13														
16														
61														
64	1	-1	-1	1	0	1	-1	-1	1	0	0	0	0	0
25,1	1	1	1	1	2	-1	-1	-1	-1	-2	2	2	-2	-2
87,1														
38,1	1	-1	1	-1	0	-1	1	-1	1	0	A	-A	-A	A
63														
14	1	-1	1	-1	0	1	-1	1	-1	0	A	-A	A	-A

SM	1	2	3	4	5
1	1	1	1	1	2
49	1	1	1	1	-2
28,1 76,1	1,T	1,T	-1,T	-1,T	0
37,1 85,1	1,T	-1,T	1,T	-1,T	0
16 64	1	-1	-1	1	0

DT	1	2	3	4	5
1	1	1	1	1	2
49	1	1	1	1	-2
3 51	1	1	-1	-1	0
26,1 74,1	1,T	-1,T	1,T	-1,T	0
28,1 76,1	1,T	-1,T	-1,T	1,T	0

M

	1	2	3	4	5
1	2	2	2	2	4
49	2	2	2	2	-4
4 52	2	2	-2	-2	0
2 3 14 15					
25,1 26,1 27,1					
28,1 38,1 39,1					
50 51 62 63					
73,1 74,1 75,1					
76,1 86,1 87,1	0	0	0	0	0
40,1 88,1	0	0	2	-2	0
37,1 85,1	0	0	-2	2	0
13 16 61 64	2	-2	0	0	0

X

	1	2	3	4
1	2	2	2	2
49	2	2	-2	-2
2 4 25,1				
26,1 28,1 50				
52 73,1 74,1				
76,1	0	0	0	0
75,1	0	0	2J	-2J
3 51	2	-2	0	0
27,1	0	0	-2J	2J

Y

	1	2	3	4	5
1	2	1	1	1	1
49	2	-1	-1	-1	-1
2	0	J	J	-J	-J
75,1	0	-1,T	1,T	-1,T	1,T
28,1	0	-1,TJ	1,TJ	1,TJ	-1,TJ
50	0	-J	-J	J	J
27,1	0	1,T	-1,T	1,T	-1,T
76,1	0	1,TJ	-1,TJ	-1,TJ	1,TJ

D

	1	2	3	4
1	1	1	1	1
49	1	1	-1	-1
28,1	1,T	-1,T	-1,TJ	1,TJ
76,1	1,T	-1,T	1,TJ	-1,TJ

GM	1+	2+	3+	4+	5+	1-	2-	3-	4-	5-	6+	7+	6-	7-
1	1	1	1	1	2	1	1	1	1	2	2	2	2	2
49	1	1	1	1	2	1	1	1	1	2	-2	-2	-2	-2
4														
52	1	1	1	1	-2	1	1	1	1	-2	0	0	0	0
2,1														
3,1														
50,1														
51,1	1	1	-1	-1	0	1	1	-1	-1	0	0	0	0	0
26,1														
27,1														
74,1														
75,1	1	1	-1	-1	0	-1	-1	1	1	0	0	0	0	0
73	1	1	1	1	2	-1	-1	-1	-1	-2	-2	-2	2	2
28														
76	1	1	1	1	-2	-1	-1	-1	-1	2	0	0	0	0
37,1														
40,1														
85,1														
88,1	1	-1	-1	1	0	-1	1	1	-1	0	0	0	0	0
39														
86	1	-1	1	-1	0	-1	1	-1	1	0	-A	A	A	-A
15														
62	1	-1	1	-1	0	1	-1	1	-1	0	-A	A	-A	A
13,1														
16,1														
61,1														
64,1	1	-1	-1	1	0	1	-1	-1	1	0	0	0	0	0
25	1	1	1	1	2	-1	-1	-1	-1	-2	2	2	-2	-2
87														
38	1	-1	1	-1	0	-1	1	-1	1	0	A	-A	-A	A
63														
14	1	-1	1	-1	0	1	-1	1	-1	0	A	-A	A	-A

SM	1	2	3	4	5
1	1	1	1	1	2
49	1	1	1	1	-2
28 76	1	1	-1	-1	0
37,1 85,1	1,T	-1,T	1,T	-1,T	0
16,1 64,1	1,T	-1,T	-1,T	1,T	0

DT	1	2	3	4	5
1	1	1	1	1	2
49	1	1	1	1	-2
3,1 51,1	1,T	1,T	-1,T	-1,T	0
26,1 74,1	1,T	-1,T	1,T	-1,T	0
28 76	1	-1	-1	1	0

M	1+	2+	3+	4+	5+	1-	2-	3-	4-	5-	6+	7+	6-	7-
1	1	1	1	1	2	1	1	1	1	2	2	2	2	2
49	1	1	1	1	2	1	1	1	1	2	-2	-2	-2	-2
4														
52	-1	-1	-1	-1	2	-1	-1	-1	-1	2	0	0	0	0
2,1														
50,1	J	J	-J	-J	0	J	J	-J	-J	0	0	0	0	0
3,1														
51,1	-J	-J	J	J	0	-J	-J	J	J	0	0	0	0	0
26,1														
74,1	J	J	-J	-J	0	-J	-J	J	J	0	0	0	0	0
27,1														
75,1	-J	-J	J	J	0	J	J	-J	-J	0	0	0	0	0
73	1	1	1	1	2	-1	-1	-1	-1	-2	-2	-2	2	2
28														
76	-1	-1	-1	-1	2	1	1	1	1	-2	0	0	0	0
40,1														
88,1	1	-1	-1	1	0	-1	1	1	-1	0	0	0	0	0
37,1														
85,1	-1	1	1	-1	0	1	-1	-1	1	0	0	0	0	0
86	J	-J	J	-J	0	-J	J	-J	J	0	AJ	-AJ	-AJ	AJ
87	-J	J	-J	J	0	J	-J	J	-J	0	AJ	-AJ	-AJ	AJ
62	J	-J	J	-J	0	J	-J	J	-J	0	AJ	-AJ	AJ	-AJ
63	-J	J	-J	J	0	-J	J	-J	J	0	AJ	-AJ	AJ	-AJ
16,1														
64,1	1	-1	-1	1	0	1	-1	-1	1	0	0	0	0	0
13,1														
61,1	-1	1	1	-1	0	-1	-1	-1	-1	0	0	0	0	0
25	1	1	1	1	2	-1	-1	-1	-1	-2	2	2	-2	-2
38	J	-J	J	-J	0	-J	J	-J	J	0	-AJ	AJ	AJ	-AJ
39	-J	J	-J	J	0	J	-J	J	-J	0	-AJ	AJ	AJ	-AJ
14	J	-J	J	-J	0	J	-J	J	-J	0	-AJ	AJ	-AJ	AJ
15	-J	J	-J	J	0	-J	J	-J	J	0	-AJ	AJ	-AJ	AJ

X

	1	2	3	4
1	2	2	2	2
49	2	2	-2	-2
4	0	0	2J	-2J
2,1 3,1 25				
26,1 27,1 50,1				
51,1 73 74,1				
75,1	0	0	0	0
28 76	-2	2	0	0
52	0	0	-2J	2J

Y

	1	2	3	4	5
1	1	1	1	1	2
49	1	1	1	1	-2
2,1 50,1	-1,TJ	-1,TJ	1,TJ	1,TJ	0
27,1 75,1	-1,TJ	1,TJ	-1,TJ	1,TJ	0
28 76	1	-1	-1	1	0

D

	1	2	3	4
1	1	1	1	1
49	1	1	-1	-1
28	1	-1	J	-J
76	1	-1	-J	J

GM	1+	2+	3+	4+	5+	1-	2-	3-	4-	5-	6+	7+	6-	7-
1	1	1	1	1	2	1	1	1	1	2	2	2	2	2
49	1	1	1	1	2	1	1	1	1	2	-2	-2	-2	-2
4														
52	1	1	1	1	-2	1	1	1	1	-2	0	0	0	0
2,1														
3,1														
50,1														
51,1	1	1	-1	-1	0	1	1	-1	-1	0	0	0	0	0
26														
27														
74														
75	1	1	-1	-1	0	-1	-1	1	1	0	0	0	0	0
73,1	1	1	1	1	2	-1	-1	-1	-1	-2	-2	-2	2	2
28,1														
76,1	1	1	1	1	-2	-1	-1	-1	-1	2	0	0	0	0
37,1														
40,1														
85,1														
88,1	1	-1	-1	1	0	-1	1	1	-1	0	0	0	0	0
39														
86	1	-1	1	-1	0	-1	1	-1	1	0	-A	A	A	-A
15,1														
62,1	1	-1	1	-1	0	1	-1	1	-1	0	-A	A	-A	A
13														
16														
61														
64	1	-1	-1	1	0	1	-1	-1	1	0	0	0	0	0
25,1	1	1	1	1	2	-1	-1	-1	-1	-2	2	2	-2	-2
87														
38	1	-1	1	-1	0	-1	1	-1	1	0	A	-A	-A	A
63,1														
14,1	1	-1	1	-1	0	1	-1	1	-1	0	A	-A	A	-A

SM	1	2	3	4	5
1	1	1	1	1	2
49	1	1	1	1	-2
28,1 76,1	1,T	1,T	-1,T	-1,T	0
37,1 85,1	1,T	-1,T	1,T	-1,T	0
16 64	1	-1	-1	1	0

DT	1	2	3	4	5
1	1	1	1	1	2
49	1	1	1	1	-2
3,1 51,1	1,T	1,T	-1,T	-1,T	0
26 74	1	-1	1	-1	0
28,1 76,1	1,T	-1,T	-1,T	1,T	0

M	1	2	3	4	5
1	2	2	2	2	4
49	2	2	2	2	-4
4 52	2	2	-2	-2	0
2,1 3,1 14,1					
15,1 25,1 26					
27 28,1 38					
39 50,1 51,1					
62,1 63,1 73,1					
74 75 76,1					
86 87	0	0	0	0	0
40,1 88,1	0	0	2	-2	0
37,1 85,1	0	0	-2	2	0
13 16 61 64	-2	2	0	0	0

X	1	2	3	4
1	2	2	2	2
49	2	2	-2	-2
3,1 4 25,1				
27 28,1 51,1				
52 73,1 75				
76,1	0	0	0	0
2,1	0	0	2J	-2J
26 74	2	-2	0	0
50,1	0	0	-2J	2J

Y	1	2	3	4	5
1	2	1	1	1	1
49	2	-1	-1	-1	-1
2,1	0	-1,T	-1,T	1,T	1,T
75	0	J	-J	J	-J
28,1	0	1,TJ	-1,TJ	-1,TJ	1,TJ
50,1	0	1,T	1,T	-1,T	-1,T
27	0	-J	J	-J	J
76,1	0	-1,TJ	1,TJ	1,TJ	-1,T.I

D	1	2	3	4
1	1	1	1	1
49	1	1	-1	-1
28,1	1,T	-1,T	-1,TJ	1,TJ
76,1	1,T	-1,T	1,TJ	-1,TJ

GROUP 65   P3

GM

	1	2	3	4	5	6
1	1	1	1	1	1	1
49	1	1	1	-1	-1	-1
3	1	W	W*	-W*	-1	-W
5	1	W*	W	W	1	W*
51	1	W	W*	W*	1	W
53	1	W*	W	-W	-1	-W*

K

	1	2	3	4	5	6
1	1	1	1	1	1	1
49	1	1	1	-1	-1	-1
3	1	W	W*	-W*	-1	-W
5	1	W*	W	W	1	W*
51	1	W	W*	W*	1	W
53	1	W*	W	-W	-1	-W*

KA

	1	2	3	4	5	6
1	1	1	1	1	1	1
49	1	1	1	-1	-1	-1
5	1	W*	W	W	1	W*
3	1	W	W*	-W*	-1	-W
53	1	W*	W	-W	-1	-W*
51	1	W	W*	W*	1	W

126

GM	1+	2+	3+	1-	2-	3-	4+	5+	6+	4-	5-	6-
1	1	1	1	1	1	1	1	1	1	1	1	1
49	1	1	1	1	1	1	-1	-1	-1	-1	-1	-1
3	1	W	W*	1	W	W*	-W*	-1	-W	-W*	-1	-W
5	1	W*	W	1	W*	W	W	1	W*	W	1	W*
13	1	1	1	-1	-1	-1	1	1	1	-1	-1	-1
15	1	W	W*	-1	-W	-W*	-W*	-1	-W	W*	1	W
17	1	W*	W	-1	-W*	-W	W	1	W*	-W	-1	-W*
51	1	W	W*	1	W	W*	W*	1	W	W*	1	W
53	1	W*	W	1	W*	W	-W	-1	-W*	-W	-1	-W*
61	1	1	1	-1	-1	-1	-1	-1	-1	1	1	1
63	1	W	W*	-1	-W	-W*	W*	1	W	-W*	-1	-W
65	1	W*	W	-1	-W*	-W	-W	-1	-W*	W	1	W*

K	1	2	3	4	5	6
1	1	1	1	1	1	1
49	1	1	1	-1	-1	-1
3	1	W	W*	-W*	-1	-W
5	1	W*	W	W	1	W*
51	1	W	W*	W*	1	W
53	1	W*	W	-W	-1	-W*

M	1+	1-	2+	2-
1	1	1	1	1
49	1	1	-1	-1
13	1	-1	1	-1
61	1	-1	-1	1

GM

	1	2	3	4	5	6
1	1	1	2	1	1	2
49	1	1	2	-1	-1	-2
3 53	1	1	-1	-1	-1	1
10 56 60	1	-1	0	J	-J	0
51 5	1	1	-1	1	1	-1
58 8 12	1	-1	0	-J	J	0

SM

	1	2	3	4
1	1	1	1	1
49	1	1	-1	-1
8	1	-1	J	-J
56	1	-1	-J	J

K

	1	2	3	4	5	6
1	1	1	1	-1	1	1
49	1	1	1	-1	-1	-1
3	1	W	W*	-W*	-1	-W
5	1	W*	W	W	1	W*
51	1	W	W*	W*	1	W
53	1	W*	W	-W	-1	-W*

M

	1	2	3	4
1	1	1	1	1
49	1	1	-1	-1
8	1	-1	J	-J
56	1	-1	-J	J

SN

	1	2	3	4
1	1	1	1	1
49	1	1	-1	-1
10	1	-1	J	-J
58	1	-1	-J	J

GM

	1	2	3	4	5	6
1	1	1	2	1	1	2
49	1	1	2	-1	-1	-2
3 53	1	1	-1	-1	-1	1
7 11 57	1	-1	0	J	-J	0
51 5	1	1	-1	1	1	-1
55 59 9	1	-1	0	-J	J	0

LD

	1	2	3	4
1	1	1	1	1
49	1	1	-1	-1
57	1	-1	J	-J
9	1	-1	-J	J

K

	1	2	3	4	5	6
1	1	1	2	1	1	2
49	1	1	2	-1	-1	-2
3 53	1	1	-1	-1	-1	1
7 11 57	1	-1	0	J	-J	0
51 5	1	1	-1	1	1	-1
55 59 9	1	-1	0	-J	J	0

M

	1	2	3	4
1	1	1	1	1
49	1	1	-1	-1
59	1	-1	J	-J
11	1	-1	-J	J

T

	1	2	3	4
1	1	1	1	1
49	1	1	-1	-1
59	1	-1	J	-J
11	1	-1	-J	J

LE

	1	2	3	4
1	1	1	1	1
49	1	1	-1	-1
7	1	-1	-J	J
55	1	-1	J	-J

KA

	1	2	3	4	5	6
1	1	1	2	1	1	2
49	1	1	2	-1	-1	-2
5 51	1	1	-1	1	1	-1
3 53	1	1	-1	-1	-1	1
9 55 59	1	-1	0	J	-J	0
7 11 57	1	-1	0	-J	J	0

TA

	1	2	3	4
1	1	1	1	1
49	1	1	-1	-1
11	1	-1	J	-J
59	1	-1	-J	J

GM

	1	2	3	4	5	6
1	1	1	2	1	1	2
49	1	1	2	-1	-1	-2
3 53	1	1	-1	-1	-1	1
19 23 69	1	-1	0	J	-J	0
51 5	1	1	-1	1	1	-1
67 71 21	1	-1	0	-J	J	0

SM

	1	2	3	4
1	1	1	1	1
49	1	1	-1	-1
71	1	-1	J	-J
23	1	-1	-J	J

K

	1	2	3	4	5	6
1	1	1	1	1	1	1
49	1	1	1	-1	-1	-1
3	1	W	W*	-W*	-1	-W
5	1	W*	W	W	1	W*
51	1	W	W*	W*	1	W
53	1	W*	W	-W	-1	-W*

M

	1	2	3	4
1	1	1	1	1
49	1	1	-1	-1
71	1	-1	J	-J
23	1	-1	-J	J

SN

	1	2	3	4
1	1	1	1	1
49	1	1	-1	-1
19	1	-1	J	-J
67	1	-1	-J	J

GM

	1	2	3	4	5	6
1	1	1	2	1	1	2
49	1	1	2	-1	-1	-2
3 53	1	1	-1	-1	-1	1
22 68 72	1	-1	0	J	-J	0
51 5	1	1	-1	1	1	-1
70 20 24	1	-1	0	-J	J	0

LD

	1	2	3	4
1	1	1	1	1
49	1	1	-1	-1
72	1	-1	J	-J
24	1	-1	-J	J

K

	1	2	3	4	5	6
1	1	1	2	1	1	2
49	1	1	2	-1	-1	-2
3 53	1	1	-1	-1	-1	1
22 68 72	1	-1	0	J	-J	0
51 5	1	1	-1	1	1	-1
70 20 24	1	-1	0	-J	J	0

M

	1	2	3	4
1	1	1	1	1
49	1	1	-1	-1
20	1	-1	J	-J
68	1	-1	-J	J

T

	1	2	3	4
1	1	1	1	1
49	1	1	-1	-1
20	1	-1	J	-J
68	1	-1	-J	J

LE

	1	2	3	4
1	1	1	1	1
49	1	1	-1	-1
22	1	-1	-J	J
70	1	-1	J	-J

	KA					
	1	2	3	4	5	6
1	1	1	2	1	1	2
49	1	1	2	-1	-1	-2
5 51	1	1	-1	1	1	-1
3 53	1	1	-1	-1	-1	1
20 24 70	1	-1	0	J	-J	0
22 68 72	1	-1	0	-J	J	0

	TA			
	1	2	3	4
1	1	1	1	1
49	1	1	-1	-1
20	1	-1	-J	J
68	1	-1	J	-J

GM

	1+	2+	3+	1-	2-	3-	4+	5+	6+	4-	5-	6-
1	1	1	2	1	1	2	1	1	2	1	1	2
49	1	1	2	1	1	2	-1	-1	-2	-1	-1	-2
3 53	1	1	-1	1	1	-1	-1	-1	1	-1	-1	1
22 68 72	1	-1	0	-1	1	0	J	-J	0	-J	J	0
10 56 60	1	-1	0	1	-1	0	J	-J	0	J	-J	0
61	1	1	2	-1	-1	-2	-1	-1	-2	1	1	2
17 63	1	1	-1	-1	-1	1	1	1	-1	-1	-1	1
51 5	1	1	-1	1	1	-1	1	1	-1	1	1	-1
70 20 24	1	-1	0	-1	1	0	-J	J	0	J	-J	0
58 8 12	1	-1	0	1	-1	0	-J	J	0	-J	J	0
13	1	1	2	-1	-1	-2	1	1	2	-1	-1	-2
65 15	1	1	-1	-1	-1	1	-1	-1	1	1	1	-1

LD

	1	2	3	4
1	1	1	1	1
49	1	1	-1	-1
72	1	-1	J	-J
24	1	-1	-J	J

SM

	1	2	3	4
1	1	1	1	1
49	1	1	-1	-1
8	1	-1	J	-J
56	1	-1	-J	J

K

	1	2	3	4	5	6
1	1	1	2	1	1	2
49	1	1	2	-1	-1	-2
3 53	1	1	-1	-1	-1	1
22 68 72	1	-1	0	J	-J	0
51 5	1	1	-1	1	1	-1
70 20 24	1	-1	0	-J	J	0

M

	1+	2+	1-	2-	3+	4+	3-	4-
1	1	1	1	1	1	1	1	1
49	1	1	1	1	-1	-1	-1	-1
20	1	-1	-1	1	J	-J	-J	J
8	1	-1	1	-1	J	-J	J	-J
61	1	1	-1	-1	-1	-1	1	1
68	1	-1	-1	1	-J	J	J	-J
56	1	-1	1	-1	-J	J	-J	J
13	1	1	-1	-1	1	1	-1	-1

T

	1	2	3	4
1	1	1	1	1
49	1	1	-1	-1
20	1	-1	J	-J
68	1	-1	-J	J

GM

	1+	2+	3+	1-	2-	3-	4+	5+	6+	4-	5-	6-
1	1	1	2	1	1	2	1	1	2	1	1	2
49	1	1	2	1	1	2	-1	-1	-2	-1	-1	-2
3 53	1	1	-1	1	1	-1	-1	-1	1	-1	-1	1
7 11 57	1	-1	0	1	-1	0	J	-J	0	J	-J	0
19 23 69	1	-1	0	-1	1	0	J	-J	0	-J	J	0
61	1	1	2	-1	-1	-2	-1	-1	-2	1	1	2
17 63	1	1	-1	-1	-1	1	1	1	-1	-1	-1	1
51 5	1	1	-1	1	1	-1	1	1	-1	1	1	-1
55 59 9	1	-1	0	1	-1	0	-J	J	0	-J	J	0
67 71 21	1	-1	0	-1	1	0	-J	J	0	J	-J	0
13	1	1	2	-1	-1	-2	1	1	2	-1	-1	-2
65 15	1	1	-1	-1	-1	1	-1	-1	1	1	1	-1

LD

	1	2	3	4
1	1	1	1	1
49	1	1	-1	-1
57	1	-1	J	-J
9	1	-1	-J	J

SM

	1	2	3	4
1	1	1	1	1
49	1	1	-1	-1
71	1	-1	J	-J
23	1	-1	-J	J

K

	1	2	3	4	5	6
1	1	1	2	1	1	2
49	1	1	2	-1	-1	-2
3 53	1	1	-1	-1	-1	1
7 11 57	1	-1	0	J	-J	0
51 5	1	1	-1	1	1	-1
55 59 9	1	-1	0	-J	J	0

M

	1+	2+	1-	2-	3+	4+	3-	4-
1	1	1	1	1	1	1	1	1
49	1	1	1	1	-1	-1	-1	-1
59	1	-1	1	-1	J	-J	J	-J
71	1	-1	-1	1	J	-J	-J	J
61	1	1	-1	-1	-1	-1	1	1
11	1	-1	1	-1	-J	J	-J	J
23	1	-1	-1	1	-J	J	J	-J
13	1	1	-1	-1	1	1	-1	-1

T

	1	2	3	4
1	1	1	1	1
49	1	1	-1	-1
59	1	-1	J	-J
11	1	-1	-J	J

GM

	1	2	3	4	5	6	7	8	9	10	11	12
1	1	1	1	1	1	1	1	1	1	1	1	1
49	1	1	1	1	1	1	-1	-1	-1	-1	-1	-1
3	1	1	W	W	W*	W*	-W*	-W*	-1	-1	-W	-W
5	1	1	W*	W*	W	W	W	W	1	1	W*	W*
4	1	-1	1	-1	1	-1	J	-J	J	-J	J	-J
6	1	-1	W	-W	W*	-W*	-U*	U*	-J	J	U	-U
50	1	-1	W*	-W*	W	-W	-U	U	J	-J	U*	-U*
51	1	1	W	W	W*	W*	W*	W*	1	1	W	W
53	1	1	W*	W*	W	W	-W	-W	-1	-1	-W*	-W*
52	1	-1	1	-1	1	-1	-J	J	-J	J	-J	J
54	1	-1	W	-W	W*	-W*	U*	-U*	J	-J	-U	U
2	1	-1	W*	-W*	W	-W	U	-U	-J	J	-U*	U*

K

	1	2	3	4	5	6
1	1	1	1	1	1	1
49	1	1	1	-1	-1	-1
3	1	W	W*	-W*	-1	-W
5	1	W*	W	W	1	W*
51	1	W	W*	W*	1	W
53	1	W*	W	-W	-1	-W*

M

	1	2	3	4
1	1	1	1	1
49	1	1	-1	-1
4	1	-1	J	-J
52	1	-1	-J	J

GM

	1	2	3	4	5	6	7	8	9	10	11	12
1	1	1	1	1	1	1	1	1	1	1	1	1
49	1	1	1	1	1	1	-1	-1	-1	-1	-1	-1
3	1	1	W	W	W*	W*	-W*	-W*	-1	-1	-W	-W
5	1	1	W*	W*	W	W	W	W	1	1	W*	W*
16	1	-1	1	-1	1	-1	J	-J	J	-J	J	-J
18	1	-1	W	-W	W*	-W*	-U*	U*	-J	J	U	-U
62	1	-1	W*	-W*	W	-W	-U	U	J	-J	U*	-U*
51	1	1	W	W	W*	W*	W*	W*	1	1	W	W
53	1	1	W*	W*	W	W	-W	-W	-1	-1	-W*	-W*
64	1	-1	1	-1	1	-1	-J	J	-J	J	-J	J
66	1	-1	W	-W	W*	-W*	U*	-U*	J	-J	-U	U
14	1	-1	W*	-W*	W	-W	U	-U	-J	J	-U*	U*

LD

	1	2	3	4
1	1	1	1	1
49	1	1	-1	-1
16	1	-1	J	-J
64	1	-1	-J	J

SM

	1	2	3	4
1	1	1	1	1
49	1	1	-1	-1
16	1	-1	J	-J
64	1	-1	-J	J

K

	1	2	3	4	5	6	7	8	9	10	11	12
1	1	1	1	1	1	1	1	1	1	1	1	1
49	1	1	1	1	1	1	-1	-1	-1	-1	-1	-1
3	1	1	W	W	W*	W*	-W*	-W*	-1	-1	-W	-W
5	1	1	W*	W*	W	W	W	W	1	1	W*	W*
16	1	-1	1	-1	1	-1	J	-J	J	-J	J	-J
18	1	-1	W	-W	W*	-W*	-U*	U*	-J	J	U	-U
62	1	-1	W*	-W*	W	-W	-U	U	J	-J	U*	-U*
51	1	1	W	W	W*	W*	W*	W*	1	1	W	W
53	1	1	W*	W*	W	W	-W	-W	-1	-1	-W*	-W*
64	1	-1	1	-1	1	-1	-J	J	-J	J	-J	J
66	1	-1	W	-W	W*	-W*	U*	-U*	J	-J	-U	U
14	1	-1	W*	-W*	W	-W	U	-U	-J	J	-U*	U*

T

	1	2	3	4
1	1	1	1	1
49	1	1	-1	-1
16	1	-1	J	-J
64	1	-1	-J	J

M

	1	2	3	4
1	1	1	1	1
49	1	1	-1	-1
16	1	-1	J	-J
64	1	-1	-J	J

B

	1	2	3	4
1	1	1	1	1
49	1	1	-1	-1
16	1	-1	J	-J
64	1	-1	-J	J

LE

	1	2	3	4
1	1	1	1	1
49	1	1	-1	-1
16	1	-1	-J	J
64	1	-1	J	-J

SN

	1	2	3	4
1	1	1	1	1
49	1	1	-1	-1
16	1	-1	-J	J
64	1	-1	J	-J

KA

	1	2	3	4	5	6	7	8	9	10	11	12
1	1	1	1	1	1	1	1	1	1	1	1	1
49	1	1	1	1	1	1	-1	-1	-1	-1	-1	-1
5	1	1	W	W	W*	W*	W*	W*	1	1	W	W
3	1	1	W*	W*	W	W	-W	-W	-1	-1	-W*	-W*
18	1	-1	W*	-W*	W	-W	-U	U	J	-J	U*	-U*
16	1	-1	1	-1	1	-1	-J	J	-J	J	-J	J
14	1	-1	W	-W	W*	-W*	U*	-U*	J	-J	-U	U
53	1	1	W	W	W*	W*	-W*	-W*	-1	-1	-W	-W
51	1	1	W*	W*	W	W	W	W	1	1	W*	W*
66	1	-1	W*	-W*	W	-W	U	-U	-J	J	-U*	U*
64	1	-1	1	-1	1	-1	J	-J	J	-J	J	-J
62	1	-1	W	-W	W*	-W*	-U*	U*	-J	J	U	-U

TA

	1	2	3	4
1	1	1	1	1
49	1	1	-1	-1
16	1	-1	-J	J
64	1	-1	J	-J

BA

	1	2	3	4
1	1	1	1	1
49	1	1	-1	-1
16	1	-1	-J	J
64	1	-1	J	-J

BB

	1	2	3	4
1	1	1	1	1
49	1	1	-1	-1
16	1	-1	-J	J
64	1	-1	J	-J

BC

	1	2	3	4
1	1	1	1	1
49	1	1	-1	-1
16	1	-1	J	-J
64	1	-1	-J	J

GROUP 75   P6/M

GM

	1+	2+	3+	4+	5+	6+	1-	2-	3-	4-	5-	6-
1	1	1	1	1	1	1	1	1	1	1	1	1
49	1	1	1	1	1	1	1	1	1	1	1	1
3	1	1	W	W	W*	W*	1	1	W	W	W*	W*
5	1	1	W*	W*	W	W	1	1	W*	W*	W	W
4	1	-1	1	-1	1	-1	1	-1	1	-1	1	-1
6	1	-1	W	-W	W*	-W*	1	-1	W	-W	W*	-W*
50	1	-1	W*	-W*	W	-W	1	-1	W*	-W*	W	-W
16	1	-1	1	-1	1	-1	-1	1	-1	1	-1	1
18	1	-1	W	-W	W*	-W*	-1	1	-W	W	-W*	W*
62	1	-1	W*	-W*	W	-W	-1	1	-W*	W*	-W	W
61	1	1	1	1	1	1	-1	-1	-1	-1	-1	-1
63	1	1	W	W	W*	W*	-1	-1	-W	-W	-W*	-W*
65	1	1	W*	W*	W	W	-1	-1	-W*	-W*	-W	-W
51	1	1	W	W	W*	W*	1	1	W	W	W*	W*
53	1	1	W*	W*	W	W	1	1	W*	W*	W	W
52	1	-1	1	-1	1	-1	1	-1	1	-1	1	-1
54	1	-1	W	-W	W*	-W*	1	-1	W	-W	W*	-W*
2	1	-1	W*	-W*	W	-W	1	-1	W*	-W*	W	-W
64	1	-1	1	-1	1	-1	-1	1	-1	1	-1	1
66	1	-1	W	-W	W*	-W*	-1	1	-W	W	-W*	W*
14	1	-1	W*	-W*	W	-W	-1	1	-W*	W*	-W	W
13	1	1	1	1	1	1	-1	-1	-1	-1	-1	-1
15	1	1	W	W	W*	W*	-1	-1	-W	-W	-W*	-W*
17	1	1	W*	W*	W	W	-1	-1	-W*	-W*	-W	-W

LD

	1	2	3	4
1	1	1	1	1
49	1	1	-1	-1
16	1	-1	J	-J
64	1	-1	-J	J

SM

	1	2	3	4
1	1	1	1	1
49	1	1	-1	-1
16	1	-1	J	-J
64	1	-1	-J	J

K

	1	2	3	4	5	6	7	8	9	10	11	12
1	1	1	1	1	1	1	-1	-1	-1	1	1	1
49	1	1	1	1	1	1	-1	-1	-1	-1	-1	-1
3	1	1	W	W	W*	W*	-W*	-W*	-1	-1	-W	-W
5	1	1	W*	W*	W	W	W	W	1	1	W*	W*
16	1	-1	1	-1	1	-1	J	-J	J	-J	J	-J
18	1	-1	W	-W	W*	-W*	-U*	U*	-J	J	U	-U
62	1	-1	W*	-W*	W	-W	-U	U	J	-J	U*	-U*
51	1	1	W	W	W*	W*	W*	W*	1	1	W	W
53	1	1	W*	W*	W	W	-W	-W	-1	-1	-W*	-W*
64	1	-1	1	-1	1	-1	-J	J	-J	J	-J	J
66	1	-1	W	-W	W*	-W*	U*	-U*	J	-J	-U	U
14	1	-1	W*	-W*	W	-W	U	-U	-J	J	-U*	U*

138

7+	8+	9+	10+	11+	12+	7-	8-	9-	10-	11-	12-
1	1	1	1	1	1	1	1	1	1	1	1
-1	-1	-1	-1	-1	-1	-1	-1	-1	-1	-1	-1
-W*	-W*	-1	-1	-W	-W	-W*	-W*	-1	-1	-W	-W
W	W	1	1	W*	W*	W	W	1	1	W*	W*
J	-J	J	-J	J	-J	J	-J	J	-J	J	-J
-U*	U*	-J	J	U	-U	-U*	U*	-J	J	U	-U
-U	U	J	-J	U*	-U*	-U	U	J	-J	U*	-U*
J	-J	J	-J	J	-J	-J	J	-J	J	-J	J
-U*	U*	-J	J	U	-U	U*	-U*	J	-J	-U	U
-U	U	-J	-J	U*	-U	U	-U	-J	J	-U	U*
-1	-1	-1	-1	-1	-1	1	1	1	1	1	1
W*	W*	1	1	W	W	-W*	-W*	-1	-1	-W	-W
-W	-W	-1	-1	-W*	-W*	W	W	1	1	W*	W*
W*	W*	1	1	W	W	W	W	1	1	W	W
-W	-W	-1	-1	-W*	-W*	-W	-W	-1	-1	-W*	-W*
-J	J	-J	J	-J	J	-J	J	-J	J	-J	J
U*	-U*	J	-J	-U	U	U*	-U*	J	-J	-U	U
U	-U	-J	J	-U*	U*	U	-U	-J	J	-U*	U*
-J	J	-J	J	-J	J	-J	J	-J	J	-J	J
U*	-U*	J	-J	-U	U	-U*	U*	-J	J	U	-U
U	-U	-J	J	-U*	U*	-U	U	J	-J	U*	-U*
1	1	1	1	1	1	-1	-1	-1	-1	-1	-1
-W*	-W*	-1	-1	-W	-W	W*	W*	1	1	W	W
W	W	1	1	W*	W*	-W	-W	-1	-1	-W*	-W*

M

	1+	2+	1-	2-	3+	4+	3-	4-
1	1	1	1	1	1	1	1	1
49	1	1	1	1	-1	-1	-1	-1
4	1	-1	1	-1	J	-J	J	-J
16	1	-1	-1	1	J	-J	-J	J
61	1	1	-1	-1	-1	-1	1	1
52	1	-1	1	-1	-J	J	-J	J
64	1	-1	-1	1	-J	J	J	-J
13	1	1	-1	-1	1	1	-1	-1

T

	1	2	3	4
1	1	1	1	1
49	1	1	-1	-1
16	1	-1	J	-J
64	1	-1	-J	J

B

	1	2	3	4
1	1	1	1	1
49	1	1	-1	-1
16	1	-1	J	-J
64	1	-1	-J	J

BA

	1	2	3	4
1	1	1	1	1
49	1	1	-1	-1
16	1	-1	-J	J
64	1	-1	J	-J

GM

	1	2	3	4	5	6	7	8	9
1	1	1	1	1	2	2	2	2	2
49	1	1	1	1	2	2	-2	-2	-2
3 53	1	1	1	1	-1	-1	1	1	-2
4 52	1	1	-1	-1	2	-2	0	0	0
6 50	1	1	-1	-1	-1	1	-B	B	0
7 9 11 55									
57 59	1	-1	1	-1	0	0	0	0	0
8 10 12 56									
58 60	1	-1	-1	1	0	0	0	0	0
51 5	1	1	1	1	-1	-1	-1	-1	2
54 2	1	1	-1	-1	-1	1	B	-B	0

LD

	1	2	3	4
1	1	1	1	1
49	1	1	-1	-1
57	1	-1	J	-J
9	1	-1	-J	J

SM

	1	2	3	4
1	1	1	1	1
49	1	1	-1	-1
56	1	-1	J	-J
8	1	-1	-J	J

K

	1	2	3	4	5	6
1	1	1	2	1	1	2
49	1	1	2	-1	-1	-2
3 53	1	1	-1	-1	-1	1
7 11 57	1	-1	0	J	-J	0
51 5	1	1	-1	1	1	-1
55 59 9	1	-1	0	-J	J	0

M

	1	2	3	4	5
1	1	1	1	1	2
49	1	1	1	1	-2
4 52	1	1	-1	-1	0
11 59	1	-1	1	-1	0
8 56	1	-1	-1	1	0

T

	1	2	3	4
1	1	1	1	1
49	1	1	-1	-1
59	1	-1	J	-J
11	1	-1	-J	J

GM

	1	2	3	4	5	6	7	8	9
1	1	1	1	1	2	2	2	2	2
49	1	1	1	1	2	2	-2	-2	-2
3 53	1	1	1	1	-1	-1	1	1	-2
4 52	1	1	-1	-1	2	-2	0	0	0
6 50	1	1	-1	-1	-1	1	-B	B	0
20 22 24 68									
70 72	1	-1	1	-1	0	0	0	0	0
19 21 23 67									
69 71	1	-1	-1	1	0	0	0	0	0
51 5	1	1	1	1	-1	-1	-1	-1	2
54 2	1	1	-1	-1	-1	1	B	-B	0

LD

	1	2	3	4
1	1	1	1	1
49	1	1	-1	-1
72	1	-1	J	-J
24	1	-1	-J	J

SM

	1	2	3	4
1	1	1	1	1
49	1	1	-1	-1
71	1	-1	J	-J
23	1	-1	-J	J

K

	1	2	3	4	5	6
1	1	1	2	1	1	2
49	1	1	2	-1	-1	-2
3 53	1	1	-1	-1	-1	1
22 68 72	1	-1	0	J	-J	0
51 5	1	1	-1	1	1	-1
70 20 24	1	-1	0	-J	J	0

M

	1	2	3	4	5
1	1	1	1	1	2
49	1	1	1	1	-2
4 52	1	1	-1	-1	0
20 68	1	-1	1	-1	0
23 71	1	-1	-1	1	0

T

	1	2	3	4
1	1	1	1	1
49	1	1	-1	-1
20	1	-1	J	-J
68	1	-1	-J	J

GM

	1	2	3	4	5	6	7	8	9
1	1	1	1	1	2	2	2	2	2
49	1	1	1	1	2	2	-2	-2	-2
3 53	1	1	1	1	-1	-1	1	1	-2
16 64	1	1	-1	-1	2	-2	0	0	0
18 62	1	1	-1	-1	-1	1	-B	B	0
19 21 23 67									
69 71	1	-1	1	-1	0	0	0	0	0
8 10 12 56									
58 60	1	-1	-1	1	0	0	0	0	0
51 5	1	1	1	1	-1	-1	-1	-1	2
66 14	1	1	-1	-1	-1	1	B	-B	0

LD

	1	2	3	4
1	1	1	1	1
49	1	1	-1	-1
16	1	-1	J	-J
64	1	-1	-J	J

SM

	1	2	3	4	5
1	1	1	1	1	2
49	1	1	1	1	-2
16 64	1	1	-1	-1	0
23 71	1	-1	1	-1	0
8 56	1	-1	-1	1	0

K

	1	2	3	4	5	6	7	8	9	10	11	12
1	1	1	1	1	1	1	1	1	1	1	1	1
49	1	1	1	1	1	1	-1	-1	1	-1	-1	-1
3	1	1	W	W	W*	W*	-W*	-W*	-1	-1	-W	-W
5	1	1	W*	W*	W	W	W	W	1	1	W*	W*
16	1	-1	1	-1	1	-1	J	-J	J	-J	J	-J
18	1	-1	W	-W	W*	-W*	-U*	U*	-J	J	U	-U
62	1	-1	W*	-W*	W	-W	-U	U	J	-J	U*	-U*
51	1	1	W	W	W*	W*	W*	W*	1	1	W	W
53	1	1	W*	W*	W	W	-W	-W	-1	-1	-W*	-W*
64	1	-1	1	-1	1	-1	-J	J	-J	J	-J	J
66	1	-1	W	-W	W*	-W*	U*	-U*	J	-J	-U	U
14	1	-1	W*	-W*	W	-W	U	-U	-J	J	-U*	U*

M

	1	2	3	4	5
1	1	1	1	1	2
49	1	1	1	1	-2
16 64	1	1	-1	-1	0
23 71	1	-1	1	-1	0
8 56	1	-1	-1	1	0

	T				
		1	2	3	4
1		1	1	1	1
49		1	1	−1	−1
16		1	−1	J	−J
64		1	−1	−J	J

	B				
		1	2	3	4
1		1	1	1	1
49		1	1	−1	−1
16		1	−1	J	−J
64		1	−1	−J	J

	SN					
		1	2	3	4	5
1		1	1	1	1	2
49		1	1	1	1	−2
10 58		1	1	−1	−1	0
16 64		1	−1	1	−1	0
19 67		1	−1	−1	1	0

	BB				
		1	2	3	4
1		1	1	1	1
49		1	1	−1	−1
16		1	−1	−J	J
64		1	−1	J	−J

GM

	1	2	3	4	5	6	7	8	9
1	1	1	1	1	2	2	2	2	2
49	1	1	1	1	2	-2	-2	-2	
3 53	1	1	1	1	-1	-1	1	1	-2
7 9 11 55									
57 59	1	1	-1	-1	0	0	0	0	0
20 22 24 68									
70 72	1	-1	1	-1	0	0	0	0	0
16 64	1	-1	-1	1	2	-2	0	0	0
18 62	1	-1	-1	1	-1	1	-B	B	0
51 5	1	1	1	1	-1	-1	-1	-1	2
66 14	1	-1	-1	1	-1	1	B	-B	0

LD

	1	2	3	4	5
1	1	1	1	1	2
49	1	1	1	1	-2
9 57	1	1	-1	-1	0
24 72	1	-1	1	-1	0
16 64	1	-1	-1	1	0

SM

	1	2	3	4
1	1	1	1	1
49	1	1	-1	-1
16	1	-1	J	-J
64	1	-1	-J	J

K

	1	2	3	4	5	6	7	8	9
1	1	1	1	1	2	2	2	2	2
49	1	1	1	1	2	2	-2	-2	-2
3 53	1	1	1	1	-1	-1	1	1	-2
7 9 11 55									
57 59	1	1	-1	-1	0	0	0	0	0
20 22 24 68									
70 72	1	-1	1	-1	0	0	0	0	0
16 64	1	-1	-1	1	2	-2	0	0	0
18 62	1	-1	-1	1	-1	1	-B	B	0
51 5	1	1	1	1	-1	-1	-1	-1	2
66 14	1	-1	-1	1	-1	1	B	-B	0

M

	1	2	3	4	5
1	1	1	1	1	2
49	1	1	1	1	-2
11 59	1	1	-1	-1	0
20 68	1	-1	1	-1	0
16 64	1	-1	-1	1	0

T

	1	2	3	4	5
1	1	1	1	1	2
49	1	1	1	1	-2
11 59	1	1	-1	-1	0
20 68	1	-1	1	-1	0
16 64	1	-1	-1	1	0

B

	1	2	3	4
1	1	1	1	1
49	1	1	-1	-1
16	1	-1	J	-J
64	1	-1	-J	J

LE

	1	2	3	4	5
1	1	1	1	1	2
49	1	1	1	1	-2
7 55	1	1	-1	-1	0
16 64	1	-1	-1	1	0
22 70	1	-1	1	-1	0

KA

	1	2	3	4	5	6	7	8	9
1	1	1	1	1	2	2	2	2	2
49	1	1	1	1	2	2	-2	-2	-2
5 51	1	1	1	1	-1	-1	-1	-1	2
3 53	1	1	1	1	-1	-1	1	1	-2
9 7 11 57									
55 59	1	1	-1	-1	0	0	0	0	0
18 62	1	-1	-1	1	-1	1	-B	B	0
16 64	1	-1	-1	1	2	-2	0	0	0
14 66	1	-1	-1	1	-1	1	B	-B	0
20 24 22 68									
72 70	1	-1	1	-1	0	0	0	0	0

TA

	1	2	3	4	5
1	1	1	1	1	2
49	1	1	1	1	-2
11 59	1	1	-1	-1	0
16 64	1	-1	-1	1	0
20 68	1	-1	1	-1	0

BA

	1	2	3	4
1	1	1	1	1
49	1	1	-1	-1
16	1	-1	-J	J
64	1	-1	J	-J

P6/M2/M2/M

GM	1+	2+	3+	4+	5+	6+	1-	2-	3-	4-	5-	6-	7+	8+	9+	7-	8-	9-
1	1	1	1	1	2	2	1	1	1	1	2	2	2	2	2	2	2	2
49	1	1	1	1	2	2	1	1	1	1	2	2	-2	-2	-2	-2	-2	-2
3																		
53	1	1	1	1	-1	-1	1	1	1	1	-1	-1	1	1	-2	1	1	-2
4																		
52	1	1	-1	-1	2	-2	1	1	-1	-1	2	-2	0	0	0	0	0	0
6																		
50	1	1	-1	-1	-1	1	1	1	-1	-1	-1	1	-B	B	0	-B	B	0
7																		
9																		
11																		
55																		
57																		
59	1	-1	1	-1	0	0	1	-1	1	-1	0	0	0	0	0	0	0	0
8																		
10																		
12																		
56																		
58																		
60	1	-1	-1	1	0	0	1	-1	-1	1	0	0	0	0	0	0	0	0
20																		
22																		
24																		
68																		
70																		
72	1	-1	-1	1	0	0	-1	1	1	-1	0	0	0	0	0	0	0	0
19																		
21																		
23																		
67																		
69																		
71	1	-1	1	-1	0	0	-1	1	-1	1	0	0	0	0	0	0	0	0
16																		
64	1	1	-1	-1	2	-2	-1	-1	1	1	-2	2	0	0	0	0	0	0
14																		
66	1	1	-1	-1	-1	1	-1	-1	1	1	1	-1	B	-B	0	-B	B	0
13	1	1	1	1	2	2	-1	-1	-1	-1	-2	-2	2	2	2	-2	-2	-2
15																		
65	1	1	1	1	-1	-1	-1	-1	-1	-1	1	1	1	1	-2	-1	-1	2
51																		
5	1	1	1	1	-1	-1	1	1	1	1	-1	-1	-1	-1	2	-1	-1	2
54																		
2	1	1	-1	-1	-1	1	1	1	-1	-1	-1	1	B	-B	0	B	-B	0
62																		
18	1	1	-1	-1	-1	1	-1	-1	1	1	1	-1	-B	B	0	B	-B	0
61	1	1	1	1	2	2	-1	-1	-1	-1	-2	-2	-2	-2	-2	2	2	2
63																		
17	1	1	1	1	-1	-1	-1	-1	-1	-1	1	1	-1	-1	2	1	1	-2

LD		1	2	3	4	5
1		1	1	1	1	2
49		1	1	1	1	-2
9	57	1	1	-1	-1	0
24	72	1	-1	1	-1	0
16	64	1	-1	-1	1	0

SM		1	2	3	4	5
1		1	1	1	1	2
49		1	1	1	1	-2
16	64	1	1	-1	-1	0
23	71	1	-1	1	-1	0
8	56	1	-1	-1	1	0

K	1	2	3	4	5	6	7	8	9
1	1	1	1	1	2	2	2	2	2
49	1	1	1	1	2	2	-2	-2	-2
3 53	1	1	1	1	-1	-1	1	1	-2
7 9 11 55									
57 59	1	1	-1	-1	0	0	0	0	0
20 22 24 68									
70 72	1	-1	1	-1	0	0	0	0	0
16 64	1	-1	-1	1	2	-2	0	0	0
14 66	1	-1	-1	1	-1	1	-B	B	0
51 5	1	1	1	1	-1	-1	-1	-1	2
62 18	1	-1	-1	1	-1	1	B	-B	0

M	1+	2+	3+	4+	1-	2-	3-	4-	5+	5-
1	1	1	1	1	1	1	1	1	2	2
49	1	1	1	1	1	1	1	1	-2	-2
4 52	1	1	-1	-1	1	1	-1	-1	0	0
11 59	1	-1	1	-1	1	-1	1	-1	0	0
8 56	1	-1	-1	1	1	-1	-1	1	0	0
20 68	1	-1	-1	1	-1	1	1	-1	0	0
23 71	1	-1	1	-1	-1	1	-1	1	0	0
16 64	1	1	-1	-1	-1	-1	1	1	0	0
13	1	1	1	1	-1	-1	-1	-1	2	-2
61	1	1	1	1	-1	-1	-1	-1	-2	2

T	1	2	3	4	5
1	1	1	1	1	2
49	1	1	1	1	-2
11 59	1	1	-1	-1	0
20 68	1	-1	1	-1	0
16 64	1	-1	-1	1	0

B	1	2	3	4
1	1	1	1	1
49	1	1	-1	-1
16	1	-1	J	-J
64	1	-1	-J	J

APPENDIX II

Compatibility Relations for the

Eighty Layer Groups

GROUP   4   PM11   GROUP   5   PB11

GM   1   2   3   4   GM   1   2   3   4
      ---------------------------------   ---------------------------------
F   1   2   3   4   F   1   2   4   3

FA   1   2   3   4   FA   1   2   4   3

A   1   2   3   4   A   1   2   3   4
      ---------------------------------   ---------------------------------
F   1   2   3   4   F   2   1   4   3

B   1   2   3   4   B   1   2   3   4
      ---------------------------------   ---------------------------------
F   1   2   3   4   F   1   2   4   3

Y   1   2   3   4   Y   1   2   3   4
      ---------------------------------   ---------------------------------
F   1   2   3   4   F   1   2   3   4

GROUP   6   P2/M11

GM   1+   2+   1-   2-   3+   4+   3-   4-

F   1   2   2   1   3   4   4   3

A   1+   2+   1-   2-   3+   4+   3-   4-

F   1   2   2   1   3   4   4   3

B   1+   2+   1-   2-   3+   4+   3-   4-

F   1   2   2   1   3   4   4   3

Y   1+   2+   1-   2-   3+   4+   3-   4-

F   1   2   2   1   3   4   4   3

GROUP 7   P2/B11

GM	1+	2+	1-	2-	3+	4+	3-	4-
F	1	2	2	1	4	3	3	4

A	1	2
F	1	3
	2	4

B	1+	2+	1-	2-	3+	4+	3-	4-
F	1	2	2	1	4	3	3	4

V	1	2
F	1	3
	2	4

GROUP 8   P112

GM	1	2	3	4
DT	1	2	3	4
TA	1	2	3	4

X	1	2	3	4
D	1	2	3	4
DA	1	2	3	4

Y	1	2	3	4
DT	1	2	3	4

S	1	2	3	4
D	1	2	3	4

GROUP 9   P112(1)

GM	1	2	3	4
DT	1	2	4	3
TA	1	2	4	3

X	1	2	3	4
D	1	2	4	3
DA	1	2	4	3

Y	1	2	3	4
DT	1	2	3	4

S	1	2	3	4
D	1	2	3	4

GROUP 10   C112

GM	1	2	3	4
DT	1	2	3	4
DA	1	2	3	4

YA	1	2	3	4
F	1	2	3	4
FA	1	2	3	4

YB	1	2	3	4
DT	1	2	3	4

GROUP 11   P11M

GM	1	2	3	4
SN	1	2	3	4
SM	1	2	3	4

X	1	2	3	4
SM	1	2	3	4

Y	1	2	3	4
C	1	2	3	4
CA	1	2	3	4

S	1	2	3	4
C	1	2	3	4

GROUP 12   P11A

GM	1	2	3	4
SN	1	2	4	3
SM	1	2	4	3

X	1	2	3	4
SM	1	2	3	4

Y	1	2	3	4
C	1	2	4	3
CA	1	2	4	3

S	1	2	3	4
C	1	2	3	4

GROUP 13   C11M

GM	1	2	3	4
SM	1	2	3	4
SN	1	2	3	4

YA	1	2	3	4
SM	1	2	3	4

YB	1	2	3	4
C	1	2	3	4
CA	1	2	3	4

GROUP 14   P112/M

GM	1+	2+	1−	2−	3+	4+	3−	4−
DT	1	2	1	2	3	4	3	4
SM	1	2	2	1	3	4	4	3

X	1+	2+	1−	2−	3+	4+	3−	4−
D	1	2	1	2	3	4	3	4
SM	1	2	2	1	3	4	4	3

Y	1+	2+	1−	2−	3+	4+	3−	4−
C	1	2	2	1	3	4	4	3
DT	1	2	1	2	3	4	3	4

S	1+	2+	1−	2−	3+	4+	3−	4−
D	1	2	1	2	3	4	3	4
C	1	2	2	1	3	4	4	3

GROUP 15   P112(1)/M

GM	1+	2+	1−	2−	3+	4+	3−	4−
DT	1	2	1	2	4	3	4	3
SM	1	2	2	1	3	4	4	3

X	1+	2+	1−	2−	3+	4+	3−	4−
D	1	2	1	2	4	3	4	3
SM	1	2	2	1	3	4	4	3

Y	1	2
C	1	3
	2	4
DT	1	3
	2	4

S	1	2
D	1	3
	2	4
C	1	3
	2	4

GROUP 16　　　　　　　　　C112/M

GM	1+	2+	1-	2-	3+	4+	3-	4-
DT	1	2	1	2	3	4	3	4
SM	1	2	2	1	3	4	4	3

YA	1+	2+	1-	2-	3+	4+	3-	4-
F	1	2	1	2	3	4	3	4
SM	1	2	2	1	3	4	4	3

YB	1+	2+	1-	2-	3+	4+	3-	4-
DT	1	2	1	2	3	4	3	4
C	1	2	2	1	3	4	4	3

GROUP 17　　　　　　　　　P112/A

GM	1+	2+	1-	2-	3+	4+	3-	4-
DT	1	2	1	2	3	4	3	4
SM	1	2	2	1	4	3	3	4

X	1	2
D	1	3
	2	4
SM	1	3
	2	4

Y	1+	2+	1-	2-	3+	4+	3-	4-
C	1	2	2	1	4	3	3	4
DT	1	2	1	2	3	4	3	4

S	1	2
D	1	3
	2	4
C	1	3
	2	4

155

GROUP 18 — P112(1)/A

GM	1+	2+	1-	2-	3+	4+	3-	4-
DT	1	2	1	2	4	3	4	3
SM	1	2	2	1	4	3	3	4

X	1	2
D	1 2	3 4
SM	1 2	3 4

Y	1	2
C	1 2	3 4
DT	1 2	3 4

S	1+	2+	1-	2-	3+	4+	3-	4-
D	1	2	1	2	3	4	3	4
C	2	1	1	2	4	3	3	4

GROUP 19 — P222

GM	1	2	3	4	5
DT	1	2	2	1	3 4
SM	1	2	1	2	3 4

X	1	2	3	4	5
D	1	2	2	1	3 4
SM	1	2	1	2	3 4

Y	1	2	3	4	5
C	1	2	1	2	3 4
DT	1	2	2	1	3 4

S	1	2	3	4	5
D	1	2	2	1	3 4
C	1	2	1	2	3 4

GROUP 20 — P222(1)

GM	1	2	3	4	5
DT	1	1	2	2	3 4
SM	1	2	2	1	3 4

X	1	2	3	4	5
D	1	1	2	2	3 4
SM	1	2	2	1	3 4

Y	1	2	3	4	5
C	1 2	4	3	3	4
DT	1 2	3	3	4	4

S	1	2	3	4	5
D	1 2	3	3	4	4
C	1 2	4	3	3	4

GROUP 21 P22(1)2(1)

GM	1	2	3	4	5
DT	1	2	2	1	3 4
SM	1	2	1	2	3 4

X	1	2	3	4	5
D	1 2	4	3	3	4
SM	1 2	3	4	3	4

Y	1	2	3	4	5
C	1 2	4	3	4	3
DT	1 2	4	3	3	4

S	1	2	3	4	5
D	1	2	2	1	3 4
C	1	2	1	2	3 4

GROUP 22 C222

GM	1	2	3	4	5
DT	1	2	2	1	3 4
SM	1	2	1	2	3 4

YA	1	2	3	4	5
F	1	2	2	1	3 4
SM	1	2	1	2	3 4

YB	1	2	3	4	5
DT	1	2	2	1	3 4
C	1	2	1	2	3 4

GROUP 23 P2MM

GM	1	2	3	4	5
DT	1	2	1	2	3 4
SM	1	2	2	1	3 4

X	1	2	3	4	5
D	1	2	1	2	3 4
SM	1	2	2	1	3 4

Y	1	2	3	4	5
C	1	2	2	1	3 4
DT	1	2	1	2	3 4

S	1	2	3	4	5
D	1	2	1	2	3 4
C	1	2	2	1	3 4

GM	1	2	3	4	5
DT	1	2	3	4	5
TA	1	2	3	4	5
SM	1	2	2	1	3 4
V	1	2	2	1	3 4
VA	1	2	2	1	3 4

DT	1	2	3	4	5
V	1	2	2	1	3 4

SM	1	2	3	4
V	1	2	4	3
VA	1	2	3	4

X	1	2	3	4	5
D	1	2	3	4	5
DA	1	2	3	4	5
SM	1	2	2	1	3 4
V	1	2	2	1	3 4
VA	1	2	2	1	3 4

V	1	2	3	4	5
C	1	2	2	1	3 4
DT	1	2	3	4	5
V	1	2	2	1	3 4

S	1	2	3	4	5
D	1	2	3	4	5
C	1	2	2	1	3 4
V	1	2	2	1	3 4

D	1	2	3	4	5
V	1	2	2	1	3 4

C	1	2	3	4
V	1	2	4	3

TA	1	2	3	4	5
VA	1	2	2	1	3 4

DA	1	2	3	4	5
VA	1	2	2	1	3 4

GM	1	2	3	4	5
DT	1	2	1	2	3 4
SN	1	2	3	4	5
SM	1	2	3	4	5
V	1	2	1	2	3 4
VB	1	2	1	2	3 4

DT	1	2	3	4
V	1	2	3	4
VB	1	2	3	4

SM	1	2	3	4	5
V	1	2	1	2	3 4

X	1	2	3	4	5
D	1	2	1	2	3 4
SM	1	2	3	4	5
V	1	2	1	2	3 4

V	1	2	3	4	5
C	1	2	3	4	5
CA	1	2	3	4	5
DT	1	2	1	2	3 4
V	1	2	1	2	3 4
VB	1	2	1	2	3 4

S	1	2	3	4	5
D	1	2	1	2	3 4
C	1	2	3	4	5
V	1	2	1	2	3 4

D	1	2	3	4
V	1	2	3	4

C	1	2	3	4	5
V	1	2	1	2	3 4

SN	1	2	3	4	5
VB	1	2	1	2	3 4

CA	1	2	3	4	5
VB	1	2	1	2	3 4

GM	1	2	3	4	5
DT	1	2	3	4	5
TA	1	2	3	4	5
SM	1	2	2	1	3 4
V	1	2	2	1	3 4
VA	1	2	2	1	3 4

DT	1	2	3	4	5
V	1	2	2	1	3 4

SM	1	2	3	4
V	1	2	3	4
VA	1	2	4	3

X	1	2	3	4	5
D	1	2	3	4	5
DA	1	2	3	4	5
SM	1	2	2	1	3 4
V	1	2	2	1	3 4
VA	1	2	2	1	3 4

Y	1	2	3	4	5
C	1	2	2	1	3 4
DT	1	2	3	4	5
V	1	2	2	1	3 4

S	1	2	3	4	5
D	1	2	3	4	5
C	1	2	2	1	3 4
V	1	2	2	1	3 4

D	1	2	3	4	5
V	1	2	2	1	3 4

C	1	2	3	4
V	1	2	4	3

DA	1	2	3	4	5
VA	1	2	2	1	3 4

TA	1	2	3	4	5
VA	1	2	2	1	3 4

GM	1	2	3	4	5
DT	1	2	3	4	5
TA	1	2	3	4	5
SM	1	2	2	1	3
					4
V	1	2	2	1	3
					4
VA	1	2	2	1	3
					4

DT	1	2	3	4	5
V	1	2	2	1	3
					4

SM	1	2	3	4
V	1	2	3	4
VA	1	2	4	3

X	1	2	3	4	5
D	1	2	3	4	5
DA	1	2	3	4	5
SM	1	2	2	1	3
					4
V	1	2	2	1	3
					4
VA	1	2	2	1	3
					4

Y	1	2	3	4	5
C	1	2	2	1	3
					4
DT	1	2	3	4	5
V	1	2	2	1	3
					4

S	1	2	3	4	5
D	1	2	3	4	5
C	1	2	2	1	3
					4
V	1	2	2	1	3
					4

D	1	2	3	4	5
V	1	2	2	1	3
					4

C	1	2	3	4
V	1	2	4	3

DA	1	2	3	4	5
VA	1	2	2	1	3
					4

TA	1	2	3	4	5
VA	1	2	2	1	3
					4

GROUP 28   P2MA

GM	1	2	3	4	5
DT	1	2	1	2	3 4
SM	1	2	2	1	3 4

Y	1	2	3	4	5
C	1	2	2	1	3 4
DT	1	2	1	2	3 4

X	1	2	3	4	5
D	1 2	3	4	3	4
SM	1 2	4	3	3	4

S	1	2	3	4	5
D	1 2	3	4	3	4
C	1 2	4	3	3	4

GROUP 29   PAM2

GM	1	2	3	4	5
DT	1	2	3	4	5
TA	1	2	3	4	5
SM	1	2	2	1	3 4
V	1	2	2	1	3 4
VA	1	2	2	1	3 4

V	1	2	3	4	5
C	1	2	2	1	3 4
DT	1	2	3	4	5
V	1	2	2	1	3 4

DT	1	2	3	4	5
V	1	2	2	1	3 4

S	1	2	3	4	5
D	1	2	3	4	5
C	1 2	4	3	3	4
V	1 2	3	4	4	3

SM	1	2	3	4
V	1	2	4	3
VA	1	2	3	4

D	1	2	3	4	5
V	1 2	3	4	4	3

X	1	2	3	4	5
D	1	2	3	4	5
DA	1	3	2	5	4
SM	1 2	4	3	3	4
V	1 2	3	4	4	3
VA	1 2	4	3	3	4

C	1	2	3	4
V	1	2	4	3

DA	1	2	3	4	5
VA	1 2	3	4	4	3

TA	1	2	3	4	5
VA	1	2	2	1	3 4

GM	1	2	3	4	5
DT	1	2	3	4	5
TA	1	2	3	4	5
SM	1	2	2	1	3 4
V	1	2	2	1	3 4
VA	1	2	2	1	3 4

DT	1	2	3	4	5
V	1	2	2	1	3 4

SM	1	2	3	4
V	1	2	4	3
VA	1	2	3	4

X	1	2	3	4	5
D	1	5	4	3	2
DA	1	4	5	2	3
SM	1 2	4	3	3	4
V	1 2	3	4	4	3
VA	1 2	4	3	3	4

Y	1	2	3	4	5
C	1	2	2	1	3 4
DT	1	2	3	4	5
V	1	2	2	1	3 4

S	1	2	3	4	5
D	1	2	3	4	5
C	1 2	4	3	3	4
V	1 2	3	4	4	3

D	1	2	3	4	5
V	1 2	3	4	4	3

C	1	2	3	4
V	1	2	4	3

DA	1	2	3	4	5
VA	1 2	3	4	4	3

TA	1	2	3	4	5
VA	1	2	2	1	3 4

GM	1	2	3	4	5
DT	1	2	3	4	5
TA	1	2	3	4	5
SM	1	2	1	2	3 4
V	1	2	1	2	3 4
VA	1	2	1	2	3 4

DT	1	2	3	4	5
V	1	2	1	2	3 4

SM	1	2	3	4
V	1	2	3	4
VA	1	2	4	3

X	1	2	3	4	5
D	1	2	3	4	5
DA	1	3	2	5	4
SM	1 2	3	4	3	4
V	1 2	3	4	3	4
VA	1 2	4	3	4	3

V	1	2	3	4	5
C	2	1	2	1	3 4
DT	1	2	3	4	5
V	1	2	1	2	3 4

S	1	2	3	4	5
D	1	2	3	4	5
C	1 2	3	4	3	4
V	1 2	3	4	3	4

D	1	2	3	4	5
V	1 2	3	4	3	4

C	1	2	3	4
V	2	1	3	4

DA	1	2	3	4	5
VA	1 2	3	4	3	4

TA	1	2	3	4	5
VA	1	2	1	2	3 4

GM	1	2	3	4	5
DT	1	2	3	4	5
TA	1	2	3	4	5
SM	1	2	2	1	3 4
V	1	2	2	1	3 4
VA	1	2	2	1	3 4

DT	1	2	3	4	5
V	1	2	2	1	3 4

SM	1	2	3	4
V	1	2	4	3
VA	1	2	3	4

X	1	2	3	4	5
D	1	4	5	2	3
DA	1	5	4	3	2
SM	1 2	3	4	4	3
V	1 2	4	3	3	4
VA	1 2	3	4	4	3

Y	1	2	3	4	5
C	2	1	1	2	3 4
DT	1	2	3	4	5
V	1	2	2	1	3 4

S	1	2	3	4	5
D	1	2	3	4	5
C	1 2	4	3	3	4
V	1 2	3	4	4	3

D	1	2	3	4	5
V	1 2	3	4	4	3

C	1	2	3	4
V	2	1	4	3

DA	1	2	3	4	5
VA	1 2	3	4	4	3

TA	1	2	3	4	5
VA	1	2	2	1	3 4

GROUP 33 — P2BA

GM	1	2	3	4	5
DT	1	2	1	2	3 4
SM	1	2	2	1	3 4

X	1	2	3	4	5
D	1 2	4	3	4	3
SM	1 2	4	3	3	4

V	1	2	3	4	5
C	1 2	4	3	3	4
DT	1 2	3	4	3	4

S	1	2	3	4	5
D	1	2	1	2	3 4
C	1	2	2	1	3 4

GROUP 34 — C2MM

GM	1	2	3	4	5
DT	1	2	1	2	3 4
SM	1	2	2	1	3 4

YA	1	2	3	4	5
F	1	2	1	2	3 4
SM	1	2	2	1	3 4

YB	1	2	3	4	5
DT	1	2	1	2	3 4
C	1	2	2	1	3 4

```
GM    1    2    3    4    5            YB    1    2    3    4    5
      --------------------------             --------------------------
DT    1    2    3    4    5            DT    1    2    3    4    5
DA    1    2    3    4    5            C     1    2    3    4    5
SM    1    2    2    1    3            P     1    2    2    1    3
                         4                                       4

P     1    2    2    1    3            S     1    2    3    4
                         4                   --------------------
                                       P     1    2    3    4
PA    1    2    2    1    3
                         4

                                       C     1    2    3    4    5
DT    1    2    3    4    5                   --------------------------
      --------------------------       P     1    2    2    1    3
P     1    2    2    1    3                                      4
                         4

                                       F     1    2    3    4    5
SM    1    2    3    4                        --------------------------
      --------------------              P     1    2    2    1    3
P     1    2    3    4                                           4
PA    1    2    4    3

                                       DA    1    2    3    4    5
YA    1    2    3    4    5                   --------------------------
      --------------------------       PA    1    2    2    1    3
F     1    2    3    4    5                                      4
FA    1    2    3    4    5
SM    1    2    2    1    3
                         4             FA    1    2    3    4    5
                                             --------------------------
P     1    2    2    1    3            PA    1    2    2    1    3
                         4                                      4

PA    1    2    2    1    3
                         4
```

GM	1	2	3	4	5
DT	1	2	3	4	5
DA	1	2	3	4	5
SM	1	2	2	1	3 4
P	1	2	2	1	3 4
PA	1	2	2	1	3 4

DT	1	2	3	4	5
P	1	2	2	1	3 4

SM	1	2	3	4
P	1	2	4	3
PA	1	2	3	4

YA	1	2	3	4	5
F	1	2	3	4	5
FA	1	2	3	4	5
SM	1	2	2	1	3 4
P	1	2	2	1	3 4
PA	1	2	2	1	3 4

YB	1	2	3	4	5
DT	2	1	4	3	5
C	1	2	2	1	3 4
P	2	1	1	2	3 4

S	1	2	3	4
P	1	2	3	4

C	1	2	3	4
P	2	1	3	4

F	1	2	3	4	5
P	1	2	2	1	3 4

DA	1	2	3	4	5
PA	1	2	2	1	3 4

FA	1	2	3	4	5
PA	1	2	2	1	3 4

P2/M2/M2/M

GM	1+	2+	3+	4+	1-	2-	3-	4-	5+	5-
DT	1	4	3	2	2	3	4	1	5	5
SM	1	4	2	3	2	3	1	4	5	5
V	1	1	2	2	2	2	1	1	3 4	3 4

DT	1	2	3	4	5
V	1	2	2	1	3 4

SM	1	2	3	4	5
V	1	2	2	1	3 4

X	1+	2+	3+	4+	1-	2-	3-	4-	5+	5-
D	1	4	3	2	2	3	4	1	5	5
SM	1	4	2	3	2	3	1	4	5	5
V	1	1	2	2	2	2	1	1	3 4	3 4

Y	1+	2+	3+	4+	1-	2-	3-	4-	5+	5-
C	1	4	2	3	2	3	1	4	5	5
DT	1	4	3	2	2	3	4	1	5	5
V	1	1	2	2	2	2	1	1	3 4	3 4

S	1+	2+	3+	4+	1-	2-	3-	4-	5+	5-
D	1	4	3	2	2	3	4	1	5	5
C	1	4	2	3	2	3	1	4	5	5
V	1	1	2	2	2	2	1	1	3 4	3 4

D	1	2	3	4	5
V	1	2	2	1	3 4

C	1	2	3	4	5
V	1	2	2	1	3 4

GM	1+	2+	3+	4+	1-	2-	3-	4-	5+	5-
DT	1	4	3	2	2	3	4	1	5	5
SM	1	2	3	4	2	1	4	3	5	5
V	1	2	1	2	2	1	2	1	3 4	3 4

DT	1	2	3	4	5
V	1	2	1	2	3 4

SM	1	2	3	4	5
V	1	2	1	2	3 4

X	1	2	3	4
D	1	1	2 5	3 4
SM	1 2	3 4	5	5
V	1 2	1 2	3 4	3 4

Y	1+	2+	3+	4+	1-	2-	3-	4-	5+	5-
C	1	2	3	4	2	1	4	3	5	5
DT	1	4	3	2	2	3	4	1	5	5
V	1	2	1	2	2	1	2	1	3 4	3 4

S	1	2	3	4
D	1	1	2 5	3 4
C	1 2	3 4	5	5
V	1 2	1 2	3 4	3 4

D	1	2	3	4	5
V	1 2	3	4	3	4

C	1	2	3	4	5
V	1	2	1	2	3 4

GM	1+	2+	3+	4+	1-	2-	3-	4-	5+	5-
DT	1	4	3	2	2	3	4	1	5	5
SM	1	4	2	3	2	3	1	4	5	5
V	1	1	2	2	2	2	1	1	3	3
									4	4

DT	1	2	3	4	5
V	1	2	2	1	3
					4

SM	1	2	3	4	5
V	1	2	2	1	3
					4

X	1	2	3	4
D	1	1	3	2
			5	4
SM	1	3	5	5
	2	4		
V	1	1	3	3
	2	2	4	4

Y	1	2	3	4
C	1	1	3	2
			5	4
DT	1	3	5	5
	2	4		
V	1	1	3	3
	2	2	4	4

S	1	2	3	4
D	1	1	2	3
			5	4
C	1	1	3	2
			4	5
V	1	1	(2)4	(2)3
	2	2		

D	1	2	3	4	5
V	1	4	3	3	4
	2				

C	1	2	3	4	5
V	1	3	4	4	3
	2				

171

GM	1+	2+	3+	4+	1-	2-	3-	4-	5+	5-
DT	1	2	3	4	2	1	4	3	5	5
SM	1	4	2	3	2	3	1	4	5	5
V	1	2	2	1	2	1	1	2	3 4	3 4

DT	1	2	3	4	5
V	1	2	2	1	3 4

SM	1	2	3	4	5
V	1	2	1	2	3 4

X	1	2	3	4
D	1 4	2 3	5	5
SM	1 3	2 4	5	5
V	(2)1	(2)2	3 4	3 4

Y	1+	2+	3+	4+	1-	2-	3-	4-	5+	5-
C	1	4	2	3	2	3	1	4	5	5
DT	1	2	3	4	2	1	4	3	5	5
V	1	2	2	1	2	1	1	2	3 4	3 4

S	1	2	3	4
D	1 4	2 3	5	5
C	1 3	2 4	5	5
V	(2)1	(2)2	3 4	3 4

D	1	2	3	4	5
V	1	2	2	1	3 4

C	1	2	3	4	5
V	1	2	1	2	3 4

GM	1+	2+	3+	4+	1-	2-	3-	4-	5+	5-
DT	1	4	3	2	2	3	4	1	5	5
SM	1	4	2	3	2	3	1	4	5	5
V	1	1	2	2	2	2	1	1	3 4	3 4

DT	1	2	3	4	5
V	1	2	2	1	3 4

SM	1	2	3	4	5
V	1	2	2	1	3 4

X	1	2	3	4
C	1	1	4 5	2 3
SM	1 3	2 4	5	5
V	1 2	1 2	3 4	3 4

Y	1+	2+	3+	4+	1-	2-	3-	4-	5+	5-
C	1	4	2	3	2	3	1	4	5	5
DT	1	4	3	2	2	3	4	1	5	5
V	1	1	2	2	2	2	1	1	3 4	3 4

S	1	2	3	4
D	1	1	4 5	2 3
C	1 3	2 4	5	5
V	1 2	1 2	3 4	3 4

D	1	2	3	4	5
V	1 2	3	4	4	3

C	1	2	3	4	5
V	1	2	2	1	3 4

GM	1+	2+	3+	4+	1-	2-	3-	4-	5+	5-
DT	1	2	3	4	2	1	4	3	5	5
SM	1	4	2	3	2	3	1	4	5	5
V	1	2	2	1	2	1	1	2	3 4	3 4

DT	1	2	3	4	5
V	1	2	2	1	3 4

SM	1	2	3	4	5
V	1	2	1	2	3 4

X	1	2	3	4
D	1	1	2 4	3 5
SM	1 2	3 4	5	5
V	1 2	1 2	3 4	3 4

Y	1	2	3	4
C	1	1	2 3	4 5
DT	1 3	2 4	5	5
V	1 2	1 2	3 4	3 4

S	1+	1-	2+	3+	4+	5+	2-	3-	4-	5-
D	1	1	2	3	4	5	3	2	5	4
C	1	1	3	4	2	5	2	5	3	4
V	1 2	1 2	3	4	4	3	4	3	3	4

D	1	2	3	4	5
V	1 2	3	4	4	3

C	1	2	3	4	5
V	1 2	4	3	4	3

GM	1+	2+	3+	4+	1-	2-	3-	4-	5+	5-
DT	1	4	2	3	2	3	1	4	5	5
SM	1	2	3	4	2	1	4	3	5	5
V	1	2	2	1	2	1	1	2	3 4	3 4

DT	1	2	3	4	5
V	1	2	1	2	3 4

SM	1	2	3	4	5
V	1	2	2	1	3 4

X	1	2	3	4
D	1	1	2 5	3 4
SM	1 2	3 4	5	5
V	1 2	1 2	3 4	3 4

V	1	2	3	4
C	1 4	2 3	5	5
DT	1 3	2 4	5	5
V	(2)1	(2)2	3 4	3 4

S	1	2	3	4
D	1	1	2 3	4 5
C	1 3	2 4	5	5
V	1 2	1 2	3 4	3 4

D	1	2	3	4	5
V	1 2	4	3	4	3

C	1	2	3	4	5
V	1	2	2	1	3 4

GM	1+	2+	3+	4+	1-	2-	3-	4-	5+	5-
DT	1	4	3	2	2	3	4	1	5	5
SM	1	4	2	3	2	3	1	4	5	5
V	1	1	2	2	2	2	1	1	3 4	3 4

DT	1	2	3	4	5
V	1	2	2	1	3 4

SM	1	2	3	4	5
V	1	2	2	1	3 4

X	1	2	3	4
D	1 4	2 3	5	5
SM	1 4	2 3	5	5
V	(2)1	(2)2	3 4	3 4

Y	1	2	3	4
C	2 3	1 4	5	5
DT	2 3	1 4	5	5
V	(2)2	(2)1	3 4	3 4

S	1+	2+	3+	4+	1-	2-	3-	4-	5+	5-
D	1	4	3	2	2	3	4	1	5	5
C	1	4	2	3	2	3	1	4	5	5
V	1	1	2	2	2	2	1	1	3 4	3 4

D	1	2	3	4	5
V	1	2	2	1	3 4

C	1	2	3	4	5
V	1	2	2	1	3 4

176

P2/A2(1)/B2(1)/M

GM	1+	2+	3+	4+	1−	2−	3−	4−	5+	5−
DT	1	2	3	4	2	1	4	3	5	5
SM	1	4	3	2	2	3	4	1	5	5
V	1	2	1	2	2	1	2	1	3 4	3 4

DT	1	2	3	4	5
V	1	2	1	2	3 4

SM	1	2	3	4	5
V	1	2	1	2	3 4

X	1	2	3	4
D	1	1	4 5	2 3
SM	2 3	1 4	5	5
V	1 2	1 2	3 4	3 4

Y	1	2	3	4
C	1 3	2 4	5	5
DT	1 3	2 4	5	5
V	(2)1	(2)2	3 4	3 4

S	1	2	3	4
D	1	1	2 5	3 4
C	3 4	1 2	5	5
V	1 2	1 2	3 4	3 4

D	1	2	3	4	5
V	1 2	3	4	3	4

C	1	2	3	4	5
V	1	2	1	2	3 4

GM	1+	2+	3+	4+	1-	2-	3-	4-	5+	5-
DT	1	4	3	2	2	3	4	1	5	5
SM	1	4	2	3	2	3	1	4	5	5
V	1	1	2	2	2	2	1	1	3 4	3 4

DT	1	2	3	4	5
V	1	2	2	1	3 4

SM	1	2	3	4	5
V	1	2	2	1	3 4

X	1	2	3	4
D	1	1	4 5	2 3
SM	1 3	2 4	5	5
V	1 2	1 2	3 4	3 4

Y	1	2	3	4
C	1	1	4 5	2 3
DT	1 3	2 4	5	5
V	1 2	1 2	3 4	3 4

S	1	2	3	4
D	1	1	2 5	3 4
C	1	1	3 4	2 5
V	1 2	1 2	(2)3	(2)4

D	1	2	3	4	5
V	1 2	3	4	4	3

C	1	2	3	4	5
V	1 2	4	3	3	4

178

GM	1+	2+	3+	4+	1-	2-	3-	4-	5+	5-
DT	1	4	3	2	2	3	4	1	5	5
SM	1	4	2	3	2	3	1	4	5	5
P	1	1	2	2	2	2	1	1	3 4	3 4

DT	1	2	3	4	5
P	1	2	2	1	3 4

SM	1	2	3	4	5
P	1	2	2	1	3 4

YA	1+	2+	3+	4+	1-	2-	3-	4-	5+	5-
F	1	4	3	2	2	3	4	1	5	5
SM	1	4	2	3	2	3	1	4	5	5
P	1	1	2	2	2	2	1	1	3 4	3 4

YB	1+	2+	3+	4+	1-	2-	3-	4-	5+	5-
DT	1	4	3	2	2	3	4	1	5	5
C	1	4	2	3	2	3	1	4	5	5
P	1	1	2	2	2	2	1	1	3 4	3 4

S	1+	2+	1-	2-	3+	4+	3-	4-
P	1	2	2	1	3	4	4	3

C	1	2	3	4	5
P	1	2	2	1	3 4

F	1	2	3	4	5
P	1	2	2	1	3 4

GM	1+	2+	3+	4+	1-	2-	3-	4-	5+	5-
DT	1	4	3	2	2	3	4	1	5	5
SM	1	4	2	3	2	3	1	4	5	5
P	1	1	2	2	2	2	1	1	3 4	3 4

DT	1	2	3	4	5
P	1	2	2	1	3 4

SM	1	2	3	4	5
P	1	2	2	1	3 4

YA	1+	2+	3+	4+	1-	2-	3-	4-	5+	5-
F	1	4	3	2	2	3	4	1	5	5
SM	2	3	1	4	1	4	2	3	5	5
P	2	2	1	1	1	1	2	2	3 4	3 4

YB	1+	2+	3+	4+	1-	2-	3-	4-	5+	5-
DT	1	4	3	2	2	3	4	1	5	5
C	1	4	2	3	2	3	1	4	5	5
P	1	1	2	2	2	2	1	1	3 4	3 4

S	1	2
P	1 2	3 4

C	1	2	3	4	5
P	1	2	2	1	3 4

F	1	2	3	4	5
P	2	1	1	2	3 4

GM	1+	2+	3+	4+	1-	2-	3-	4-	5+	6+	7+	8+	5-	6-	7-	8-
DT	1	1	2	2	2	2	1	1	3	3	4	4	4	4	3	3
SM	1	1	2	2	2	2	1	1	3	3	4	4	4	4	3	3
D	1	1	2	2	2	2	1	1	3	3	4	4	4	4	3	3
DA	1	1	2	2	2	2	1	1	4	4	3	3	3	3	4	4

SM	1	2	3	4
D	1	2	3	4
DA	1	2	4	3

DT	1	2	3	4
D	1	2	3	4

M	1+	2+	3+	4+	1-	2-	3-	4-	5+	6+	7+	8+	5-	6-	7-	8-
SM	1	1	2	2	2	2	1	1	3	3	4	4	4	4	3	3
Y	1	1	2	2	2	2	1	1	3	3	4	4	4	4	3	3
DA	1	1	2	2	2	2	1	1	4	4	3	3	3	3	4	4
D	1	1	2	2	2	2	1	1	3	3	4	4	4	4	3	3

X	1+	2+	1-	2-	3+	4+	3-	4-
Y	1	2	2	1	3	4	4	3
DT	1	2	2	1	3	4	4	3
D	1	2	2	1	3	4	4	3

Y	1	2	3	4
D	1	2	3	4

P4/N

GM	1+	2+	3+	4+	1-	2-	3-	4-	5+	6+	7+	8+	5-	6-	7-	8-
DT	1	1	2	2	2	2	1	1	4	4	3	3	3	3	4	4
SM	1	1	2	2	2	2	1	1	4	4	3	3	3	3	4	4
D	1	1	2	2	2	2	1	1	4	4	3	3	3	3	4	4
DA	1	1	2	2	2	2	1	1	3	3	4	4	4	4	3	3

SM	1	2	3	4
D	1	2	3	4
DA	1	2	4	3

DT	1	2	3	4
D	1	2	3	4

M	1	2	3	4
SM	1	1	3	3
	2	2	4	4
V	1	1	3	3
	2	2	4	4
DA	1	1	3	3
	2	2	4	4
D	1	1	3	3
	2	2	4	4

X	1	2
Y	1	3
	2	4
DT	1	3
	2	4
D	1	3
	2	4

V	1	2	3	4
D	2	1	3	4

GROUP 53   P422

GM	1	2	3	4	5	6	7
DT	1	1	2	2	1 2	3 4	3 4
SM	1	2	2	1	1 2	3 4	3 4

M	1	2	3	4	5	6	7
SM	1	2	2	1	1 2	3 4	3 4
Y	1	1	2	2	1 2	3 4	3 4

X	1	2	3	4	5
Y	1	2	1	2	3 4
DT	1	2	2	1	3 4

GROUP 54   P42(1)2

GM	1	2	3	4	5	6	7
DT	1	1	2	2	1 2	3 4	3 4
SM	1	2	2	1	1 2	3 4	3 4

M	1	2	3	4	5	6	7
SM	1	2	2	1	1 2	3 4	3 4
Y	1	1	2	2	1 2	3 4	3 4

X	1	2	3	4	5
Y	1 2	4	3	4	3
DT	1 2	4	3	3	4

GROUP 55 P4MM

GM	1	2	3	4	5	6	7
DT	1	1	2	2	1 2	3 4	3 4
SM	1	2	1	2	1 2	3 4	3 4

M	1	2	3	4	5	6	7
SM	1	2	1	2	1 2	3 4	3 4
Y	1	1	2	2	1 2	3 4	3 4

X	1	2	3	4	5
Y	1	2	2	1	3 4
DT	1	2	1	2	3 4

GROUP 56 P4BM

GM	1	2	3	4	5	6	7
DT	1	1	2	2	1 2	3 4	3 4
SM	1	2	1	2	1 2	3 4	3 4

M	1	2	3	4	5	6	7
SM	1	2	1	2	1 2	3 4	3 4
Y	2	2	1	1	1 2	3 4	3 4

X	1	2	3	4	5
Y	1 2	4	3	3	4
DT	1 2	3	4	3	4

GROUP 57 — P4-2M

GM	1	2	3	4	5	6	7
DT	1	1	2	2	1 2	3 4	3 4
SM	1	2	1	2	1 2	3 4	3 4

M	1	2	3	4	5	6	7
SM	1	2	1	2	1 2	3 4	3 4
Y	1	1	2	2	1 2	3 4	3 4

X	1	2	3	4	5
Y	1	2	1	2	3 4
DT	1	2	2	1	3 4

GROUP 58 — P4-2(1)M

GM	1	2	3	4	5	6	7
DT	1	1	2	2	1 2	3 4	3 4
SM	1	2	1	2	1 2	3 4	3 4

M	1	2	3	4	5	6	7
SM	1	2	1	2	1 2	3 4	3 4
Y	1	1	2	2	1 2	3 4	3 4

X	1	2	3	4	5
Y	1 2	4	3	4	3
DT	1 2	4	3	3	4

GROUP 59 — P4-M2

GM	1	2	3	4	5	6	7
DT	1	1	2	2	1 2	3 4	3 4
SM	1	2	2	1	1 2	3 4	3 4

M	1	2	3	4	5	6	7
SM	1	2	2	1	1 2	3 4	3 4
Y	1	1	2	2	1 2	3 4	3 4

X	1	2	3	4	5
Y	1	2	2	1	3 4
DT	1	2	1	2	3 4

GROUP 60 — P4-B2

GM	1	2	3	4	5	6	7
DT	1	1	2	2	1 2	3 4	3 4
SM	1	2	2	1	1 2	3 4	3 4

M	1	2	3	4	5	6	7
SM	2	1	1	2	1 2	3 4	3 4
Y	2	2	1	1	1 2	3 4	3 4

X	1	2	3	4	5
Y	1 2	4	3	3	4
DT	1 2	3	4	3	4

GM	1+	2+	3+	4+	5+	1-	2-	3-	4-	5-	6+	7+	6-	7-
DT	1	1	4	4	2 3	2	2	3	3	1 4	5	5	5	5
SM	1	2	2	1	3 4	4	3	3	4	1 2	5	5	5	5
D	1	1	1	1	(2)2	2	2	2	2	(2)1	3 4	3 4	3 4	3 4

SM	1	2	3	4	5
D	1	1	2	2	3 4

DT	1	2	3	4	5
D	1	2	2	1	3 4

M	1+	2+	3+	4+	5+	1-	2-	3-	4-	5-	6+	7+	6-	7-
SM	1	2	2	1	3 4	4	3	3	4	1 2	5	5	5	5
Y	1	1	4	4	2 3	2	2	3	3	1 4	5	5	5	5
D	1	1	1	1	(2)2	2	2	2	2	(2)1	3 4	3 4	3 4	3 4

X	1+	2+	3+	4+	1-	2-	3-	4-	5+	5-
Y	1	4	2	3	2	3	1	4	5	5
DT	1	4	3	2	2	3	4	1	5	5
D	1	1	2	2	2	2	1	1	3 4	3 4

Y	1	2	3	4	5
D	1	2	2	1	3 4

GM	1+	2+	3+	4+	5+	1-	2-	3-	4-	5-	6+	7+	6-	7-
DT	1	1	4	4	2 3	2	2	3	3	1 4	5	5	5	5
SM	1	2	2	1	3 4	4	3	3	4	1 2	5	5	5	5
D	1	1	1	1	(2)2	2	2	2	2	(2)1	3 4	3 4	3 4	3 4

SM	1	2	3	4	5
D	1	1	2	2	3 4

DT	1	2	3	4	5
D	1	2	2	1	3 4

M	1	2	3	4	5
SM	1 4	2 3	1 3	2 4	(2)5
Y	1	1	1	1	2 3 4 5
D	1 2	1 2	1 2	1 2	(2)3 (2)4

X	1	2	3	4
Y	1	1	3 5	2 4
DT	1 2	3 4	5	5
D	1 2	1 2	3 4	3 4

Y	1	2	3	4	5
D	1 2	3	4	4	3

P4/M2(1)/B2/M

GM	1+	2+	3+	4+	5+	1-	2-	3-	4-	5-	6+	7+	6-	7-
DT	1	1	4	4	2 3	2	2	3	3	1 4	5	5	5	5
SM	1	2	2	1	3 4	4	3	3	4	1 2	5	5	5	5
D	1	1	1	1	(2)2	2	2	2	2	(2)1	3 4	3 4	3 4	3 4

SM	1	2	3	4	5
D	1	1	2	2	3 4

DT	1	2	3	4	5
D	1	2	2	1	3 4

M	1+	2+	3+	4+	5+	1-	2-	3-	4-	5-	6+	7+	6-	7-
SM	3	4	4	3	1 2	2	1	1	2	3 4	5	5	5	5
Y	2	2	3	3	1 4	1	1	4	4	2 3	5	5	5	5
D	2	2	2	2	(2)1	1	1	1	1	(2)2	3 4	3 4	3 4	3 4

X	1	2	3	4
Y	2 3	1 4	5	5
DT	2 3	1 4	5	5
D	(2)2	(2)1	3 4	3 4

Y	1	2	3	4	5
D	1	2	2	1	3 4

GM	1+	2+	3+	4+	5+	1-	2-	3-	4-	5-	6+	7+	6-	7-
DT	1	1	4	4	2 3	2	2	3	3	1 4	5	5	5	5
SM	1	2	2	1	3 4	4	3	3	4	1 2	5	5	5	5
D	1	1	1	1	(2)2	2	2	2	2	(2)1	3 4	3 4	3 4	3 4

SM	1	2	3	4	5
D	1	1	2	2	3 4

DT	1	2	3	4	5
D	1	2	2	1	3 4

M	1	2	3	4	5
SM	2 3	1 4	1 3	2 4	(2)5
Y	1	1	1	1	2 3 4 5
D	1 2	1 2	1 2	1 2	(2)3 (2)4

X	1	2	3	4
Y	1	1	4 5	2 3
DT	1 3	2 4	5	5
D	1 2	1 2	3 4	3 4

Y	1	2	3	4	5
D	1 2	4	3	3	4

GROUP 67   P312

GM	1	2	3	4	5	6
SN	1	2	1 2	3	4	3 4
SM	1	2	1 2	4	3	3 4

M	1	2	3	4
SM	1	2	3	4

GROUP 68   P321

GM	1	2	3	4	5	6
LD	1	2	1 2	3	4	3 4
LE	1	2	1 2	4	3	3 4

K	1	2	3	4	5	6
LD	1	2	1 2	3	4	3 4
T	1	2	1 2	4	3	3 4

M	1	2	3	4
T	1	2	3	4
TA	1	2	4	3

KA	1	2	3	4	5	6
TA	1	2	1 2	4	3	3 4
LE	1	2	1 2	3	4	3 4

GROUP 69 — P3M1

GM	1	2	3	4	5	6
SN	1	2	1 2	3	4	3 4
SM	1	2	1 2	4	3	3 4

M	1	2	3	4
SM	1	2	3	4

GROUP 70 — P31M

GM	1	2	3	4	5	6
LD	1	2	1 2	3	4	3 4
LE	1	2	1 2	4	3	3 4

K	1	2	3	4	5	6
LD	1	2	1 2	3	4	3 4
T	1	2	1 2	4	3	3 4

M	1	2	3	4
T	1	2	3	4
TA	1	2	4	3

KA	1	2	3	4	5	6
TA	1	2	1 2	4	3	3 4
LE	1	2	1 2	3	4	3 4

GROUP 71   P3-12/M

GM	1+	2+	3+	1-	2-	3-	4+	5+	6+	4-	5-	6-
LD	1	2	1 2	2	1	1 2	3	4	3 4	4	3	3 4
SM	1	2	1 2	1	2	1 2	4	3	3 4	4	3	3 4

K	1	2	3	4	5	6
LD	1	2	1 2	3	4	3 4
T	1	2	1 2	4	3	3 4

M	1+	2+	1-	2-	3+	4+	3-	4-
T	1	2	2	1	3	4	4	3
SM	1	2	1	2	3	4	3	4

GROUP 72   P3-2/M1

GM	1+	2+	3+	1-	2-	3-	4+	5+	6+	4-	5-	6-
LD	1	2	1 2	1	2	1 2	3	4	3 4	3	4	3 4
SM	1	2	1 2	2	1	1 2	4	3	3 4	3	4	3 4

K	1	2	3	4	5	6
LD	1	2	1 2	3	4	3 4
T	1	2	1 2	4	3	3 4

M	1+	2+	1-	2-	3+	4+	3-	4-
T	1	2	1	2	3	4	3	4
SM	1	2	2	1	3	4	4	3

GM	1	2	3	4	5	6	7	8	9	10	11	12
SN	1	2	1	2	1	2	4	3	4	3	4	3
LD	1	2	1	2	1	2	3	4	3	4	3	4
SM	1	2	1	2	1	2	3	4	3	4	3	4
LE	1	2	1	2	1	2	4	3	4	3	4	3
B	1	2	1	2	1	2	3	4	3	4	3	4
BA	1	2	1	2	1	2	4	3	4	3	4	3
BB	1	2	1	2	1	2	4	3	4	3	4	3
BC	1	2	1	2	1	2	3	4	3	4	3	4

LD	1	2	3	4
B	1	2	3	4
BB	1	2	4	3

SM	1	2	3	4
B	1	2	3	4
BA	1	2	4	3

K	1	2	3	4	5	6	7	8	9	10	11	12
LD	1	2	1	2	1	2	3	4	3	4	3	4
T	1	2	1	2	1	2	3	4	3	4	3	4
B	1	2	1	2	1	2	3	4	3	4	3	4
BB	1	2	1	2	1	2	4	3	4	3	4	3

M	1	2	3	4
T	1	2	3	4
TA	1	2	4	3
SM	1	2	3	4
B	1	2	3	4
BA	1	2	4	3

T	1	2	3	4
B	1	2	3	4

LE	1	2	3	4
BA	1	2	3	4

SN	1	2	3	4
BB	1	2	3	4
BC	1	2	4	3

KA	1	2	3	4	5	6	7	8	9	10	11	12
TA	1	2	1	2	1	2	3	4	3	4	3	4
LE	1	2	1	2	1	2	3	4	3	4	3	4
BA	1	2	1	2	1	2	3	4	3	4	3	4

TA	1	2	3	4
BA	1	2	3	4

GM	1+	2+	3+	4+	5+	6+	1-	2-	3-	4-	5-	6-	7+	8+	9+	10+
LD	1	2	1	2	1	2	2	1	2	1	2	1	3	4	3	4
SM	1	2	1	2	1	2	2	1	2	1	2	1	3	4	3	4
B	1	2	1	2	1	2	2	1	2	1	2	1	3	4	3	4
BA	1	2	1	2	1	2	2	1	2	1	2	1	4	3	4	3

11+	12+	7-	8-	9-	10-	11-	12-
3	4	4	3	4	3	4	3
3	4	4	3	4	3	4	3
3	4	4	3	4	3	4	3
4	3	3	4	3	4	3	4

LD	1	2	3	4
B	1	2	3	4

SM	1	2	3	4
B	1	2	3	4
BA	1	2	4	3

K	1	2	3	4	5	6	7	8	9	10	11	12
LD	1	2	1	2	1	2	3	4	3	4	3	4
T	1	2	1	2	1	2	3	4	3	4	3	4
B	1	2	1	2	1	2	3	4	3	4	3	4

M	1+	2+	1-	2-	3+	4+	3-	4-
T	1	2	2	1	3	4	4	3
SM	1	2	2	1	3	4	4	3
B	1	2	2	1	3	4	4	3
BA	1	2	2	1	4	3	3	4

T	1	2	3	4
B	1	2	3	4

GROUP 76   P622

GM	1	2	3	4	5	6	7	8	9
LD	1	2	1	2	1 2	1 2	3 4	3 4	3 4
SM	1	2	2	1	1 2	1 2	3 4	3 4	3 4

K	1	2	3	4	5	6
LD	1	2	1 2	3	4	3 4
T	I	2	1 2	4	3	3 4

M	1	2	3	4	5
T	1	2	1	2	3 4
SM	1	2	2	1	3 4

GROUP 77   P6MM

GM	1	2	3	4	5	6	7	8	9
LD	1	2	1	2	1 2	1 2	3 4	3 4	3 4
SM	1	2	2	1	1 2	1 2	3 4	3 4	3 4

K	1	2	3	4	5	6
LD	1	2	1 2	3	4	3 4
T	1	2	1 2	4	3	3 4

M	1	2	3	4	5
T	1	2	1	2	3 4
SM	1	2	2	1	3 4

GM	1	2	3	4	5	6	7	8	9
SN	1	3	4	2	1 3	2 4	5	5	5
LD	1	1	2	2	(2)1	(2)2	3 4	3 4	3 4
SM	1	2	3	4	1 2	3 4	5	5	5
B	1	1	2	2	(2)1	(2)2	3 4	3 4	3 4
BB	1	1	2	2	(2)1	(2)2	3 4	3 4	3 4

LD	1	2	3	4
B	1	2	3	4
BB	1	2	4	3

SM	1	2	3	4	5
B	1	1	2	2	3 4

K	1	2	3	4	5	6	7	8	9	10	11	12
LD	1	2	1	2	1	2	3	4	3	4	3	4
T	1	2	1	2	1	2	3	4	3	4	3	4
B	1	2	1	2	1	2	3	4	3	4	3	4
BB	1	2	1	2	1	2	4	3	4	3	4	?

M	1	2	3	4	5
T	1	1	2	2	3 4
SM	1	2	3	4	5
B	1	1	2	2	3 4

T	1	2	3	4
B	1	2	3	4

SN	1	2	3	4	5
BB	1	2	1	2	3 4

GM	1	2	3	4	5	6	7	8	9
LD	1	2	3	4	1 4	2 3	5	5	5
SM	1	2	2	1	(2)1	(2)2	3 4	3 4	3 4
LE	1	2	3	4	1 4	2 3	5	5	5
B	1	2	2	1	(2)1	(2)2	3 4	3 4	3 4
BA	1	2	2	1	(2)1	(2)2	3 4	3 4	3 4

LD	1	2	3	4	5
B	1	2	2	1	3 4

SM	1	2	3	4
B	1	2	3	4
BA	1	2	4	3

K	1	2	3	4	5	6	7	8	9
LD	1	2	3	4	1 4	2 3	5	5	5
T	1	2	3	4	1 4	2 3	5	5	5
B	1	2	2	1	(2)1	(2)2	3 4	3 4	3 4

M	1	2	3	4	5
T	1	2	3	4	5
TA	1	2	3	4	5
SM	1	2	2	1	3 4
B	1	2	2	1	3 4
BA	1	2	2	1	3 4

T	1	2	3	4	5
B	1	2	2	1	3 4

LE	1	2	3	4	5
BA	1	2	2	1	3 4

KA	1	2	3	4	5	6	7	8	9
TA	1	2	3	4	1 4	2 3	5	5	5
LE	1	2	3	4	1 4	2 3	5	5	5
BA	1	2	2	1	(2)1	(2)2	3 4	3 4	3 4

TA	1	2	3	4	5
BA	1	2	2	1	3 4

GM	1+	2+	3+	4+	5+	6+	1-	2-	3-	4-	5-	6-
LD	1	4	2	3	1 4	2 3	2	3	1	4	2 3	1 4
SM	1	2	3	4	1 2	3 4	4	3	2	1	3 4	1 2
B	1	1	2	2	(2)1	(2)2	2	2	1	1	(2)2	(2)1

	7+	8+	9+	7-	8-	9-
	5	5	5	5	5	5
	5	5	5	5	5	5
	3 4	3 4	3 4	3 4	3 4	3 4

LD	1	2	3	4	5
B	1	2	2 ·	1	3 4

SM	1	2	3	4	5
B	1	1	2	2	3 4

K	1	2	3	4	5	6	7	8	9
LD	1	2	3	4	1 4	2 3	5	5	5
T	1	2	3	4	1 4	2 3	5	5	5
B	1	2	2	1	(2)1	(2)2	3 4	3 4	3 4

M	1+	2+	3+	4+	1-	2-	3-	4-	5+	5-
T	1	4	2	3	2	3	1	4	5	5
SM	1	2	3	4	4	3	2	1	5	5
B	1	1	2	2	2	2	1	1	3 4	3 4

T	1	2	3	4	5
B	1	2	2	1	3 4